EXPLORING CAREERS IN BROADCAST JOURNALISM

EXPLORING CAREERS IN BROADCAST JOURNALISM

by
ROD VAHL

The Rosen Publishing Group
New York

Published in 1983 by The Rosen Publishing Group, Inc.
29 East 21st Street, New York, N.Y. 10010

First Edition

Manufactured in the United States of America

Library of Congress Cataloging in Publication Data

Vahl, Rod.
Exploring careers in broadcast journalism.

Summary: Presents guidance, direction, and
motivation for those considering a career in
broadcast journalism, based on information from
people actually involved in the field.
1. Broadcast journalism—Vocational guidance.
2. Journalists—United States—Interviews.
[1. Broadcast journalism. 2. Journalism.
3. Occupations] I. Title.
PN4784.B75V34 1983 070.1'9'02373 82-23119
ISBN 0-8239-0595-0

To all those ambitious teens
who have made the *Blackhawk*
what it is

ABOUT THE AUTHOR

Rod Vahl is journalism teacher and adviser for the *Blackhawk* newspaper at Central High School, Davenport, Iowa. He is the author of *The Student Journalist and Consumer Reporting,* published by Richards Rosen Press, and of *Effective Editorial Writing,* a booklet published by the Quill and Scroll Society for high school journalists. He is a contributing editor for two journalism magazines, *Quill and Scroll Magazine* and *Communication: Journalism Education Today* of the Journalism Education Association. He has also written more than 200 articles for professional journals and popular magazines.

In 1981 Vahl was honored with three awards: the Medal of Merit by the Journalism Education Association, the Pioneer Award by the National Scholastic Press Association, and the Indiana Scholastic Journalism Award by Ball State University. Among his numerous other awards are the Gold Key Award by Columbia Scholastic Press Association, the Kenneth Stratton Award for Distinguished Service to Scholastic Journalism by the Iowa High School Press Association, and the Distinguished Achievement Award by the Educational Press Association of America.

Vahl has taught or spoken at numerous scholastic journalism conventions and workshops in many states, including Illinois, Iowa, Missouri, Nebraska, Minnesota, Arkansas, Georgia, Wyoming, California, and New York. He is currently the director of a high school summer journalism workshop at Ball State University.

As adviser for the *Blackhawk,* Vahl has seen his writers and photographers achieve many individual state and national awards. The newspaper continues to be a nationally ranked publication, having earned the George H. Gallup Award in 1982 for the eighth consecutive year.

ACKNOWLEDGMENTS

No doubt any venture into writing is an education for the author. Certainly that is true of this author in the writing of this book. And there were many "teachers" who deserve a special word of gratitude for sharing their experiences, expertise, and interest. No, not "teachers" in the tradition of the classroom; rather, teachers "in the field"—those who really know the realities of broadcast journalism.

Those teachers are Jane Pauley of the "Today" show on NBC-TV; Joan Lunden of "Good Morning, America" on ABC-TV; Peter Sturtevant, national news editor of CBS-TV; Rick Brown, CBS-TV news bureau chief in Chicago; John Drury, anchorman on WGN-TV in Chicago; Forrest Respess, news producer on WGN-TV; Milo Hamilton, television sportscaster for the Chicago Cubs and Chicago Bulls; Dan Miller, executive producer of public affairs for the Iowa Public Broadcasting Network; Michele Bille, news director of WQAD-TV, Moline, Illinois; Carol Clark, news reporter for WQAD-TV; and Steve Olson, news reporter and radio news anchorman for WHBF radio and television in Rock Island, Illinois.

There was another group of "teachers"—the many organizations that offered the author innumerable materials. Among them were the Radio and Television News Directors Association, the Association of Independent Television Stations, the Associated Press Broadcast Services, the National Association of Broadcasters, the National Public Radio, and the Public Broadcasting Service. And, indeed, the public relations officials of ABC, CBS, and NBC, who opened the doors for personal interviews.

Special thanks for many favors must be extended to Dick Johns, executive secretary of the Quill and Scroll Society; Diane Roberts and Eileen Butenschoen for help in typing the manuscript; Ruth Rosen of the Rosen Publishing Group for guidance and direction; my wife, Dorothy, for personal belief, and my children, Jerry and Jane, for confidence; and to all the staff members of the *Blackhawk* for their tolerance of their adviser's personal ambitions and, above all, for their concern and respect.

CONTENTS

	Introduction	1
I.	You Must Have a Love for News	3
II.	It's More than a Desire for Money	10
III.	There Are Jobs for Every Interest	16
IV.	Steve Olson—"It gets into your blood!"	26
V.	Carol Clark—"My first story—it was terrible!"	32
VI.	Adlai Stevenson—"A reporter's most valuable asset is his common sense."	38
VII.	Jane Pauley—"In an interview, I'm the least important."	43
VIII.	Peter Sturtevant—"I like to know things first."	47
IX.	Rick Brown—"I'm always competing against two other guys."	50
X.	Forrest Respess—"I've got to have a little pizzazz!"	54
XI.	Joan Lunden—"I learned and made my own mistakes right on the air."	58
XII.	John Drury—"You have to have a regard for people."	63
XIII.	Milo Hamilton—"I felt the whole world was waiting to hear from me!"	68
XIV.	Dan Miller—"I was always around people who always taught you to think."	73
XV.	An Educator Looks at Broadcasting	79
XVI.	What They Want from You	87
XVII.	Which Journalism School to Choose?	92
	Appendices	101
	Recommended Reading	121

INTRODUCTION

This author vividly recalls many articles, brochures, and books written about the teaching profession. So often the career guidance literature was researched and prepared by the professor—the man in the ivory tower. There seemed to be so much theory and so little of the practical. And as this teacher experienced his first days in the classroom, he soon found himself turning to other teachers for guidance, direction, and motivation.

That is precisely the primary direction of this book. The bulk of the information and advice comes from those who are actually involved in broadcast journalism—from the big names on network television in New York City to the reporter on Main Street in Moline, Illinois. All of them have college degrees; all are still working to fulfill their lifelong aspirations; all "tell it like it is."

The reader will discover both similarities and differences in their attitudes and ideas. But the one point that all have in common is the feeling of excitement in broadcast journalism.

The intent of this book is to permit those personalities to share that excitement with students who aspire to a career in broadcast journalism. Those personalities, more than any set of statistics or list of guidelines, will help the reader become aware of what it takes to voice that news story over radio and television.

Chapter I

YOU MUST HAVE A LOVE FOR NEWS

Walter Cronkite once said to a reporter, "Television—it's a tube and through it you get the world in your room."

That is precisely the task of the broadcast journalist—to take the news of the world into more than 80,000,000 homes in the United States. That news ranges from what is happening on the streets of villages and towns in America to the political maneuverings on Capitol Hill and in the White House, to the bloody battlefields of Afghanistan.

That news is sometimes earthshaking, sometimes frivolous. But always it tells the viewer of something that is happening to people somewhere in the totality of the world. And that is precisely the task of the broadcast journalist—to go out into the world and get that story.

It is no wonder that young people are attracted to broadcast journalism, because it is the reporter who walks the streets, stalks the halls, and crawls the battlefields to record both the facts and the emotions of the daily dramas experienced by all people. It's the horrible destruction of homes and lives in a ghetto fire. It's the rescue of a stranded Angora cat atop a flooded garage. It's the glamorous, star-studded Oscar awards. It's the smoke-filled hotel rooms during a political convention. It's the brutal, hard-hitting defensive battle of the Sugar Bowl.

The reporter gets out to that action. The reporter interviews. The reporter writes the story. The reporter delivers that story on the air. That's broadcast journalism.

NBC's Jessica Savitch was only fifteen years old in high school in Margate, New Jersey, when she first heard her voice on a high school radio show. She told a reporter, "It was mystical. I just knew I wanted to cover news events. I'd always been a writer—I wrote poems and stories as a child. It seemed the ideal thing; it all fitted together."

Somewhere along the line in every aspiring journalist's life there is that one moment, that one experience, that one resolve—"I want to cover the news!" It may emerge from simply watching local and national reporters doing their jobs on television. It may emerge from the joy of writing stories, poems, and essays. It may emerge from writing for the high school or college newspaper. But no matter the source, there seems to be one common element in all journalists—that constant desire to be involved in the breaking news of the day.

With over 10,000 radio and television stations in the nation and the number still growing, there are opportunities for those seeking a career in broadcast journalism. Vernon A. Stone, who wrote "Careers in Broadcast News" for the Radio-Television News Directors Association, provides an overall view:

> First, radio and television are the most powerful existing news media. The camera and microphone transmit actuality—the sights and sounds of events as they happen, of history being made. We witness armed conflict in a distant nation at suppertime. We and millions of others watch and listen as world leaders are interviewed by Ted Koppel or Catherine Mackin. The realities of ghettos and terrorism fill our screens. We follow the progress of a tornado or power blackout via transistor radio. Never before have there been such complete and immediate instruments for reporting.
>
> Second, the veterans in the profession truly do welcome bright, imaginative young men and women. Broadcasting itself is young, almost an infant compared to print outlets—newspapers and magazines. Approaches to radio and television news are everchanging. They have had less time than newspapers to become set in their ways. There's more premium on innovation, less on making your work fit a standardized format. Furthermore, radio and television news themselves are too demanding and too competitive to permit stagnation.
>
> Newcomers with the education and personal characteristics needed can usually find jobs in broadcast journalism if they are willing to start with realistic entry-level positions. An estimated 30,000 persons were working in radio and TV news by the early 1980s. Turnover is high enough that several hundred of these are beginners at any time. So, jobs are there. But it is a competitive field, and aspirants quite frequently fail to get or keep one of these jobs. This is often because they made a poor career choice.

Characteristics of the Broadcast Journalist

The news reporter, anchorman, director, producer—every person in that newsroom—have characteristics in common. The degree to which the prospective broadcast journalist possesses these characteristics will determine not only his or her survival in a most competitive field, but also his or her progress toward personal goals. The following list of characteristics certainly is not all-inclusive, but as the product of many interviews with every kind of newsroom personnel both on local and network stations, it can well provide some insight:

1. A deep love for news: wanting to observe what is occurring in every dimension of life and to report it to the world.

2. An intense interest in current events: a reporter will be assigned to cover every kind of news event, from chasing an ambulance to an interstate highway four-car crash to covering a school board meeting that merely elects a president.
3. An appreciation and knowledge of history, economics, political science, government, and business. Much of the broadcast journalist's work focuses upon government, its actions, and how it affects the daily lives of people.
4. A love and knowledge of writing skills. Every word uttered on a newscast is written by a reporter, an editor, an anchorman, a producer—and this writing demands special skills in broadcast journalism.
5. A curious nature, especially for discovering the unknown. Often a reporter must conduct extensive probes by interviewing witnesses and officials to find the total story. The reporter must be curious about the HOW and WHY as well as the simple facts.
6. An aggressive approach in "getting the story." Few stories come into the newsroom. The reporter must have an aggressive nature in wanting to find out the story behind a simple news tip.
7. A willingness to do research. The reporter often must spend hours making telephone calls or scouring the libraries to background a story or series of stories.
8. A social awareness; that is, an interest in and curiosity about all people. The reporter must branch out from his own personal environment and be willing to venture into all social circles, all socioeconomic groups to get the story.
9. A competitive spirit. Television news must submit itself to the ratings system, and competition is a dominant reality in broadcast journalism. Other stations and newspapers are after the same story. Exclusives are wanted. And because several other stations and newspapers are covering the same story, the reporter needs a special angle to his or her own story.
10. An imaginative mind. The reporter must not rely solely upon the assignment editor or news director to provide assignments. The reporter is expected to come up with his own ideas, to note the national story and imagine how it could be localized to his own specific audience.
11. Avidity for reading—both local and national newspapers, magazines, and books.
12. An empathy for all people—that is, the ability to project oneself into the lives, the feelings, and the thoughts of others.
13. An attractive appearance—no, not glamorous. Rather, a pleasing appearance of hairstyle, dress, smile, and manner.
14. A pleasing voice. The television reporter does *write* the story, but he or she adds another dimension—*speaking* the story. It is im-

portant to deliver the story in a manner that is both pleasing to and understandable by the viewer.

15. A desire to travel. The reporter must travel in all kinds of weather to all kinds of places. At the network correspondent level, it means traveling throughout the nation and the world. That sounds exciting, but all reporters will confess that such travel is hard on home life.
16. A willingness to work at all hours. A reporter cannot be a clock watcher if he or she wants to progress. The hours in broadcast journalism include anytime, anywhere, anyplace. Indeed, many a reporter has worked overtime without compensation, especially in the first years.
17. Ambition for advancement—ambition to have the story first, to have a better story than the other station or even his own colleagues, to move upward to the major stations and networks.
18. Constant self-evaluation. No one moves forward without improvement of oneself in writing, in interviewing, in speaking. Progress in broadcast journalism is a constant improvement in these skills.
19. Education—that means a college degree and constant reading to expand upon what one has already learned.
20. Alertness to all that is around one, awareness of new developments and trends, of movements and actions that are not yet news events.

How To Start

High school: Right now is the time to start preparing for a career in broadcast journalism, and there are several activities and attitudes one should develop immediately:

1. Take your studies seriously. Every subject you take will help to prepare you for your next formal step in education, college.
2. Work on the school newspaper or yearbook to gain journalistic writing skills. If your school has any connection with a radio or television station, enter that program in some way.
3. If your school offers any kind of journalism or photography courses, enroll in them for formal training. Also, take speech courses and don't shy away from acting in school plays.
4. View regularly the news and special affairs programs on both commercial and public television. Listen to the words, the voices, the entire program. Be critical.
5. With a tape recorder, practice delivering news. Have your tapes evaluated by your speech teachers or a local radio or television newsperson.

6. Be a station rat. That's the high school student who hangs around the radio and television stations as much as is permitted. Learn as much as you can.
7. Read biographies of broadcast journalists, both in magazines and books.
8. Read newspapers and news magazines as a daily routine. Study how the stories are written. Look for points of style that you yourself can use at this time.
9. Become active in school and community organizations, projects, and drives. Strive for leadership positions.
10. Inquire early about colleges and universities that offer journalism programs. Also, investigate scholarship programs.

College. A college degree is practically a "must" for achieving a broadcast journalism career, but there is wide disagreement as to the value of a major in broadcast journalism. The television industry is a relatively young field, and although many colleges and universities offer a broadcast journalism degree, the current consensus is that such a degree is not necessarily the best route to follow.

There are at least three characteristics of broadcast journalism that influence this lack of strong support for a broadcast journalism major:

1. Most news reporting focuses upon governmental, social, economic, and business events, both on local and national levels. It becomes extremely important that broadcast reporters have a strong educational background in the liberal arts, preferably with a concentration upon those particular areas.
2. Although not always to the pleasure of the purebred journalist, broadcast journalism has a large element of "show business" in that the physical attractiveness, the personality, and the voice of the reporter are of more influence upon the viewer than the story itself.
3. Writing is important, but more vital to the success of the broadcast journalist will be the scope of the reporter's knowledge of current events and issues. Electronic journalism is often branded as "headline journalism," and the hard-nosed, in-depth writer will not find it to his liking.

When news directors speak of the new reporter, they usually say, "I'm not so concerned that he knows much about how a television station operates. What's more important to me is that he is an aggressive newsperson who can go out and get the most important aspects of a story and tell that story in thirty or sixty seconds. There is just no time for the TV reporter to sit down and write a lengthy story."

Thus, most broadcast journalists recommend one of the following programs of college study:

1. A major in economics, history, government, political science, or a related area, with a minor in journalism (particularly broadcast journalism).
2. A double major of study, with one in broadcast journalism and the other in economics, history, government, and so on.
3. A major in broadcast journalism, but with strong electives in economics, history, government, and so on.

A popular recommendation is to earn a bachelor of arts degree in a nonjournalism major such as economics, and then to earn a master's degree in journalism or mass communications.

Practically every broadcast journalist will say, "Don't go to college to learn to edit videotape. You can quickly learn to do all the technical work simply by watching and asking questions. What you need to learn is how government functions, how business works, and how all of the social and economic forces affect people's lives."

However, colleges and universities are responding quickly to the opinions of the professional broadcast journalist, and programs are constantly changing to meet the particular needs of the field. Students should visit and study closely those colleges offering majors in broadcast journalism. They should scrutinize the curriculum to assure that it permits them a wide latitude of elective courses from which to select their four-year program.

An additional advantage are the radio and television facilities the college or university offers students to gain practical experience. Many have closed-circuit cable television stations and student-operated radio stations. Others offer some work on local radio and television stations.

In summary, talk with professional broadcast journalists, visit colleges, and determine which will best meet your needs to prepare for a professional career.

Internship. Many radio and television stations cooperate with colleges and universities in offering internships. These are periods of time during which a broadcast journalism student works in a station, performing most of the duties of an employee. Some stations pay a small wage or salary; others pay nothing. But it is extremely valuable to obtain an internship in order to gain experience.

First Job. No one starts with the anchor chair on the national network. No one is named a network correspondent as a starting job. And very, very few are lucky enough to grab a job with a major market. The first step for practically everyone is the small or medium-sized television–radio market. That's where one learns the most. In the smaller markets, the reporter does everything—reporting, filming, editing, scripting, announcing, anchoring. That's the real training for a career.

Chapter II

IT'S MORE THAN A DESIRE FOR MONEY

If you want to place yourself before the American public, the best route is through radio or television. The Radio Advertising Bureau estimates that 78,600,000 households in the nation have at least one radio, and the A. C. Nielsen Company estimates that 77,800,000 households have at least one television set.

That's a mighty high number of homes—in fact, 99 percent of homes for radio and 98 percent for television. It is no wonder that every person or group imaginable turns to television to convey what they consider an important message to the American public. Billions of dollars are spent by advertisers to persuade the public that their products and services are the best. Thousands of actors, writers, producers, and directors exert their every creative and imaginative effort to entertain the viewers. Politicians spend millions of dollars during each campaign to persuade the American voters that they can best represent them on the local, state, and national levels of government.

Television alone reached a record advertising revenue in 1979, with earnings of $9,181,900,000. While newspapers and magazines slump in advertising revenue, the television industry has increased its advertising earnings each year of its existence. Add approximately $3,000,000,000 that radio stations earn, and one soon recognizes that radio-television is a more than $12 billion industry annually—and still growing.

As of January, 1981, there were 1,019 commercial television stations operating in the United States. The AM radio stations numbered 4,575, and FM stations numbered 4,358. Add several hundred educational television network stations throughout the nation, and one realizes that over 10,000 radio and television stations are reaching into the homes of Americans.

Nearly 200,000 persons are employed full- and part-time on these radio and television stations, and more young people today are looking toward electronic journalism as a potential career.

Approximately one-third of those jobs are in broadcast journalism, and they range from the beginning street reporter on a small radio–television station in the Northwest to the top news executive on the network station. In between those positions are a good number of positions that offer the toughest of challenges and competitive drives leading one from that obscure station to the coveted role as anchorperson on the network evening newscast. Reporter, scriptwriter, editor, assignment editor, director, producer, sportscaster, weatherman, editorial specialist, anchorman, foreign correspondent, morning-show co-host, national news editor, vice president of news. There are more.

And each job demands a number of skills that must be developed, cultivated, and honed if one has any desire to progress. It is a hard-working profession, filled with sharp-edged competiton from both fellow workers and competing station personnel.

But it is also an exciting profession, with each day offering a new experience that demands the best of reportorial skills, speech delivery, and personal appearance. Indeed, it is a blend of journalism and show business, but as one journalist said, "The most exciting job—you're always where the action is!"

Every news department has its own philosophy concerning news coverage, and on the first day of work many reporters are given a handbook or manual to guide them in their news reporting labors.

It is well worth one's time to read such a manual or handbook. The following excerpts are from the handbook of WQAD-TV, an ABC affiliate station in Moline, Illinois.

THE ACTIVE EIGHT NEWS HANDBOOK

Philosophy: Nothing better explains what the Active Eight News is striving to achieve than our promotional campaign, "We're on Your Side." This should always be reflected in the kinds of stories we cover and the way we cover them. Your role is that of *surrogate viewer*. You are the eyes and ears of our audience and your responsibility is to inform them of important events and people in our community that affect our lives.

When you sit down and begin working on a story, the first question you should ask yourself is:

"Why am I doing this story?"

If you can't answer this question effectively in a sentence or two, it means one of two things . . . either there is no logical reason to do the report or you haven't thought it out well enough to develop a strong angle. If you *can* answer the question, there are still several things to consider before proceeding:

1. Why is this story important?
2. How does this story affect people?
3. How can I best visualize the story?

4. How can I make the story interesting to those who ordinarily have no interest in it?
5. How can I make my report distinctive?

When you get into the actual process of gathering information, writing and voicing your story, there are several other important points to remember:

1. Take an aggressive attitude. Remember, you are the eyes and ears of the viewer. It is up to you to ask the same tough questions they are going to ask if a situation seriously affects them. For example, what questions would *you* ask if a landfill suddenly appeared next to your property? What facts would you seek out if your company was asking you to take a considerable pay cut to keep your job?
2. All reporting should be of an explanatory nature. The viewer should be able to easily understand or comprehend anything you say. A convenient question to ask yourself is:

   "If I were sitting at home listening to this story, could I easily understand the background and complexities of the issue or situation?"

3. Copy should be delivered in a down-to-earth, conversational manner. If the copy isn't conversational, rewrite it.
4. Demonstrate considerable enthusiasm in your report. After all, if you are not seen as interested in the report, why should the viewer?

Finally, there are a few other considerations—Does this story lend itself to live coverage? Can this story be told more completely in a two- or three-part report?

Using different interviews or soundbites and a different standup, is there a way to construct this story so that it can easily be made different for separate newcasts? Can this story be tied in with the magazine show? Can the 6 PM report be teased and promoted with an on-set appearance at 5 PM possibly augmented with an in-studio guest?

Can this story be followed up? A common misconception is that because a story has been covered once or covered by competing media in the market, the story is dead. Wrong! A good story, especially one broken by us, should be worked thoroughly. Remember: that strong story you are so proud of has been seen by only a small minority of our audience the first time around.

Seriously consider going through the checklist on these three pages every time you sit down to work out a story. I guarantee it will save you a lot of time in the long run and lead to strong, effective reports.

Consider that professional pilots with thousands of hours of flight time go through a detailed checklist everytime before takeoff—it saves them more than time!

Salaries

The broadcast journalist usually starts his or her career at a low salary on a low-market radio or television station. As the journalist progresses in his or her career, the salary increases, but significantly only if the journalist moves to higher positions of responsibility or to the major markets.

The following study of television–radio salaries for broadcast journalists was made by the Radio and Television News Directors Association in 1981.

TABLE 1: Weekly News Salaries in 1981

	Television		Radio	
	Median	Mean	Median	Mean
Staff low	$200	$223	$200	$214
Staff high	$530	$654	$250	$312
5-year veteran	$350	$361	$275	$304
Cameraperson	$250	$282	—	—
Top reporter	$330	$381	$239	$268
Top anchor	$500	$623	$250	$303
News director	$500	$566	$250	$298

Note: Mean salaries are the average dollar amounts. Median salaries are midpoints of salary distribution. A few stations in most market-sized categories pay much higher than others. For that reason, medians are generally the "typical" values.

The prospective broadcast journalist must recognize that the "big money" lies in the major markets, as can be noted in Table 2, a study also made by the Radio and Television News Directors Association in 1981.

For example, a top reporter in the smallest television markets earned an average salary of only $266 per week, but that jumps to $600 per week in the top twenty-five television markets (that is, the twenty-five largest television stations in the nation in terms of advertising revenue—ADI).

The top-paying position is the anchorperson, who earned an average of $375 weekly in the smallest markets but reached $1,200 a week in the major markets. Consult Table 2 for the salaries of other positions.

The national news directors' organization best summarizes the salary possibilities in radio and television broadcast journalism in a booklet entitled "Careers in Broadcast News." It reads:

> Salaries in broadcast news vary widely with the type of station, the size of market it serves and—very important—what the individual has to offer.
>
> Able, imaginative, dependable radio and TV journalists who give their jobs that "little extra" which spells quality are always in demand and generally at attractive salaries.
>
> But do not count on becoming a fabulously paid star. The highly publicized top network salaries of up to a million dollars a year apply only to very few persons. For every John Chancellor or Barbara Walters, there are dozens of local anchorpersons making $15,000 to $25,000 a year.
>
> Modest pay is to be expected during those first years when you are gaining experience and developing your talents. No matter how well you have done in school, you'll become a really valuable professional only through experience on a full-time job.
>
> Starting salaries in radio and TV news are comparable in most areas to those paid by daily newspapers. Top salaries tend to be higher for television than for print media.
>
> Broadcast news salaries generally kept pace with cost-of-living increases during the 1970s. National surveys conducted for RTNDA across the decade showed salaries going up an average of 5–10 percent annually, roughly parallel to the rate of inflation.
>
> In the early 1980s, the lower-paid TV and radio newspersons were averaging about $12,000 a year in large cities and down to $9,000 or so in smaller markets. The cost of living normally is less, of course, away from the big cities.
>
> For the highest-paid members of news staffs—news directors and leading anchorpersons in most cases—differences by locale were greater. In the 50 largest markets in the United States top salaries averaged around $39,000 a year for TV and $27,000 for radio. In smaller markets, top pay averaged about $19,000 for TV and $11,000 for radio. Small radio stations often have only one person working in news.
>
> At the networks and many stations in major cities, most salaries are based on agreements with such unions as the Writers Guild and the American Federation of Television and Radio Artists. There is often a "talent fee" premium for appearing on camera or microphone.

TABLE 2: TV and Radio News Weekly Salaries by Market Size and Region

	Staff low		Staff high		5-year vet		Top reporter		Top anchor		News director	
Television	Median	Mean	Median	Mean	Median	Mean	Median	Mean	Median	Mean	Median	Mean
ADI 1-25	$275	$267	$962	$1218	$480	$492	$600	$649	$1200	$1296	$770	$790
ADI 26-50	$232	$252	$814	$893	$375	$400	$422	$477	$850	$903	$754	$773
ADI 51-100	$200	$224	$575	$611	$350	$348	$360	$363	$535	$581	$551	$567
ADI 101-150	$200	$202	$480	$500	$300	$324	$300	$302	$400	$439	$481	$494
ADI 151-214	$190	$212	$435	$461	$300	$332	$266	$298	$375	$398	$430	$453
East	$200	$238	$476	$567	$307	$364	$351	$379	$497	$578	$497	$530
South	$191	$200	$550	$612	$310	$327	$315	$344	$476	$563	$540	$566
Midwest	$210	$230	$539	$731	$350	$376	$330	$399	$480	$675	$501	$565
West	$225	$240	$500	$632	$400	$389	$360	$403	$500	$645	$500	$579
Radio												
Major mkts.	$275	$294	$503	$495	$400	$426	$385	$385	$499	$462	$526	$504
Large mkts.	$200	$222	$301	$346	$300	$354	$251	$287	$280	$326	$300	$349
Medium mkts.	$200	$201	$260	$297	$250	$261	$225	$248	$250	$267	$250	$278
Small mkts.	$180	$186	$211	$230	$229	$236	$200	$213	$210	$262	$210	$226
East	$180	$208	$251	$310	$268	$311	$226	$265	$248	$314	$250	$306
South	$200	$207	$250	$281	$249	$287	$201	$243	$248	$301	$235	$271
Midwest	$200	$207	$250	$312	$275	$294	$250	$269	$250	$288	$250	$298
West	$200	$237	$282	$349	$302	$331	$250	$303	$250	$319	$250	$318

Chapter III

THERE ARE JOBS FOR EVERY INTEREST

The news reporting function of television and radio is rapidly expanding at both the local and network level, with many kinds of job opportunities available to the person seeking a career in television and radio. There certainly is the allure of travel, meeting people, and reporting the events of the world. But there are many responsibilities that need to be recognized before one considers such a career. It is a competitive field, with much hiring and firing at both levels, as may be witnessed simply by viewing the daily newscasts emanating from both local and national studios.

Just what is the job? Let us visit KLXK in Middletown, America, and have some of the newspeople speak for themselves.

News Director

Come on in, I'm Jeff Miller, the news director of KLXK here in Middletown. Yes, I'm sorta' the boss here—at least here in the newsroom. My job? Well, it's much like any other news director's job in television. It's my task to hire reporters, editors, cameramen, producers, directors, scriptwriters, and anyone else who is involved in the news process. Then it's up to me to decide what all these people will do as part of their jobs. This varies a lot from station to station, depending upon size. The larger the station, the larger the staff and the responsibilities. For instance, in the very large stations, you might find the sports department a separate department, but usually a news director will have much to do with that, too.

Believe me, the news director's job is really a twenty-four-hour job because the breaking news recognizes no hours. I've got to be alert to what's going on everywhere every hour of the day. When news breaks, there are a lot of decisions to be made, and I've got to be able to do that. Sometimes that means I have to be up in the middle of the night and moving myself and others into action to cover that news. In fact, in some tight situations, I've got to be ready to fill in myself, and I might admit that I don't mind doing it at times just to remind myself what my staff must go through to get those stories on the newscast.

How do you get up this high? Well, first of all, you start right at the bottom where everyone else does—out on the street as a reporter. You've got to work your way up through most of the positions that pertain to the broadcast. I don't mean you have to have years of experience as an anchorman, but you need the experience of the reporter, the producer, the director. I guess I'd say you ought to have some experience in every aspect of newscast production.

If at all possible, get some experience in high school—especially on the school newspaper, where you quickly become acquainted with news writing, editing, and meeting deadlines. If there's any chance at all to work with television or radio, do that, too. Any kind of communications experience will help you. And don't forget to take all the writing courses you possibly can in high school.

There is a lot of discussion as to just what you should take in college. I would strongly suggest some kind of major that includes journalism courses and a lot of courses in government, economics, business, law. I say that simply because that is what you will be dealing with most of the time in broadcast journalism. You're out there covering city hall, the county courthouse, and the state legislatures, and you need to know just what is going on and why. Some colleges offer a broadcasting major, and I think that's great if you still can work in a lot of those other courses. The best approach is a broad liberal arts background. And, of course, if there is a college radio or television station, you should work there as much as possible.

Remember my saying the news director's job is a twenty-four-hour job? Well, I think I could say that is true for anyone in the broadcasting business. You've just got to be alert to everything that's going on.

I think you'll find most news directors have some special characteristics—such as ambition, energy, and eagerness to succeed. In this job you have to make a lot of quick decisions about news—decisions about what should be covered, how you want the reporters to cover the news, and the philosophy of the department as to just what is news. And you've got to be able to be critical of your reporters in a constructive way. You can't forget that they want to succeed, too, and they need all the advice and counsel you can give them. In this job you still are a reporter, but there is that added dimension—being an administrator. You've got to be able to organize your staff, provide them guidelines, implement procedures, and then evaluate what you are doing.

It's exciting—especially in this job where you have the chance to call the shots. But don't forget, although I am the one who has to fire a reporter, I can get fired too!

Producer–Director

Yes, I'm Sarah Biggers, the producer–director of our newscasts, and I'll try to explain my job. I say try because the producer's and director's job varies from station to station. Sometimes the duties are split between the two positions, and sometimes they are combined. Again, a lot depends upon the size of the station. But the job basically involves coordinating all those aspects in getting out the newscast. I usually work closely with the news director, especially at the start of the day, and then it's my job to see that all the tapes, film, videotapes, scripts, and anything else get into one neat package to put over the air.

To get this done, I am constantly in touch with the reporters, editors, soundmen, and all the others in the station who have anything to do with getting out the newscast. And, because I'm also director, I take care of the actual broadcast from the director's booth here. It is my final responsibility as to what goes on the air, so I need to be closely involved in the final preparation of the script. I work closely with the anchorpeople who do the broadcast and also with the weather broadcaster and the sports news broadcaster.

Yes, you need to gain some experience in reporting, writing, and editing to move to this kind of a job in broadcast journalism. You really need to know how to be a team member, too, because you work closely with all those people I've mentioned. Much like the news director, you must be able to make quick decisions and adjust to them immediately. For instance, it may be only a couple of hours before broadcast and you suddenly learn that there is a major fire in the downtown business district. Well, you go into action at once to get as much coverage as possible. That takes some decision-making abilities and some organizational skills. In fact, I've seen the time when we've changed the news content right during the newscast.

Besides that experience, you need some college training, and, as others tell you, some courses in broadcast journalism are preferred. I really recommend a journalism major, though that isn't the only way to get a job in TV broadcast journalism. There's so much you can learn right here working in the station, but the better prepared you are when you come in, the faster you can move.

I think the thrill in my particular job as a producer and director is to be the person who puts the whole newscast together and sees it on the screen as you planned it. It's an excitement unmatched when everything goes as you planned.

News Reporter

So you want to get into TV news. Well, I'm Larry Miller, news reporter, one of about eight reporters on this station. I suppose the best way I can tell you about my job is simply to say that I'm the person who goes where the action is. And that's where we all start. Oh, I really don't want to be a producer, director, or top executive—at least, not for a long time. What I really want to do is to become a network correspondent. Can you imagine that? Traveling all over the country and all over the world, chasing stories? Well, I want that, but for now I'm chasing the stories right here in Middletown, and that's plenty of action for now. Every day I go somewhere different. Yes, I usually get around to City Hall every day because that's my regular beat. But that's not all I do, and you might well see me chasing some ambulance to a four-car crash on the interstate or covering a hot textbook censorship case at a school board meeting. And, once in a while you get those nice fluffy assignments like covering the county fair beauty pageant. Yeah, I prefer the hard news, but those little features are a good change of pace.

The best duty is when I get to do a whole package of one or two minutes—that is, I do the story, the voicing, the whole bit. Two important things you need—all those reporting skills and good speech delivery. You've got to go out with your cameramen, do the interviewing, go back and write the story, and often help the editor with editing the story for broadcast.

But that's not all there is. Just like those competitors down at the newspaper, you spend a lot of hours sitting at your desk making phone calls or getting over to some library and doing research for a story. You still have to know what to ask when you go out on the story. That's why it's so important for the TV reporter to be really aware and alert to what is going on. You're constantly reading newspapers and magazines, so you'll be as well prepared as possible for a story assignment that is given to you only a few hours before it goes on the air.

You'll probably start out in news reporting in a small market station, and your job will include quite a bit more. You may be lugging around the camera and tape recorder yourself, and you just might have to do all the editing, scripting, and broadcasting yourself. But don't worry, you'll want to know everything there is to know about this business if you have any plans to move up to the bigger television markets.

I was a good writer when I was in high school, and I worked on the high school newspaper for three years, working up to editor in chief. I really wasn't certain that I wanted to go into journalism, so I majored in political science and took a minor in journalism. I think that's the best route, because most of my news reporting centers on governmental units such as the city council, school boards, crime commission, and county government affairs.

I hope you're inquisitive, because that's one of the most important qualities of a news reporter. You just can't take everything you're told at face value. You've got to dig and probe into many of your stories if you're going to get the whole story. So don't be afraid to be curious.

And I should add aggressive. The competition can be mighty tough if there are other stations, and don't forget you're also competing with the daily newspaper. You'll try to get an exclusive story, but most of the time the other station and newspaper reporters will be after the same story. So you've got to be aggressive and imaginative so that your story isn't the same as theirs. You'll want to have your own angle, your own treatment.

And that finally brings it all to style. I think any good reporter will develop a style—a style to his writing and to his delivery over the air. And that's just as exciting as going out into the field and getting the interview.

It's you—your story, your voice. That's neat!

Newscaster

The newscaster? Yes, that's me, Brian Downing. I guess you could call me the anchorman, too, because I'm the one responsible for reading most of the news or leading in to other reporters' stories. It's my task to see that all the preparation behind the scenes is presented in a pleasing way. That means, I guess, that I have to broadcast the news with a good voice that appeals to a wide audience, that I must have an attractive appearance on the screen, and that I must convey the news stories in a clear, understandable, and appealing manner.

That's not easy, and I'm always criticizing myself—looking at what I said and how I said it. Am I projecting my voice so that it is understood and is appealing? And it's a precarious spot, because of the ratings that you must fight in this business. If our station is running poorly in the ratings, the top brass will surely be looking at me to evaluate what I'm doing. And they might just decide that someone else can do it better.

As a journalist, a reporter, you hate that show business aspect of broadcasting, but it's there and you've got to accept it or get out. The anchorpersons are the key personalities on the newscast, and they've got to come across with the public. That's why we are often referred to as the "talent" rather than as newscaster or journalist. You're competing with other anchorpeople. You ask people which of the network stations they listen to, and a lot of them will say "Rather" or "Brokaw"—they might not even know the call letters of the network or station. They just go by "Rather" or "Pauley" or whoever.

You prepare for this job just like all the others—via the reporting route. Oh, you'll find some spots that go to a nonjournalism personality sometimes, especially in sports and specialized areas. But generally speaking, you come up through the reporting ranks, which means you've got to have the newswriting, reporting, and editing skills. In fact, in the smaller markets, the anchorman will be doing reporting out in the field as well.

I know I do, and I like that. I wouldn't want to just sit and read the news.

It isn't a matter of working a half-hour day. If you're not out on a story, you'll be researching and interviewing for special stories you want to do yourself. And you've got to be aware of what is going on in the world just as well as any reporter or news director. You're really more than a personality if you're good in this business.

Take the network anchormen—they've all been out in the field as reporters on smaller stations, then on to the networks as correspondents. And don't forget all those events such as the political conventions. The anchormen are right there, not only telling you what is going on, but also offering commentary and interviewing politicians. You can't do that without a lot of background.

You come up through the ranks after college. You can't go anywhere in broadcasting without the professional academic training. Yes, you can major in broadcast journalism, but you'll find most of the top people in this business have majored in other areas such as economics, political science, business, law. I think the ideal would be a double major, in journalism and something like economics.

I just want to add one more thing—I remember our journalism teacher in high school who gave us just three words to remember—*observe, judge, act*. Get out there and see what's really going on, decide how you should cover it, and get back and write your story.

I wouldn't want to do any other kind of work than this.

Sportscaster

You're right! You heard the reporters call me Jocko. But they're just jealous of me. My name is Tom Ensdorf, and I'm the sportscaster. I double as sports director, too. Basically, I do the sport news on both of the evening news programs, and I do the play-by-play for our radio station. Once in a while we do a telecast of a local or state athletic team, and I get to do that.

Yeah, I'd like to be a network sportscaster someday, but it's tough to get up there. But I love sports enough that I'll be happy to stay here awhile and hopefully get up to a major market or to a major league baseball town where I could do the play-by-play of a major league team.

Yes, I'm the only one in this department, and that means a lot of work. In fact, often I have to do the videotaping of a game on Friday or Saturday night. That's really rough. But I think what I like is that I have those four or five minutes of my own on the newscast. I get to do the planning, writing, and broadcasting of it all. And sports is my bag.

It helps if you've had athletic experience. I was in three sports in high school and one in college. Oh, you don't have to have it, but you need that intense interest in and knowledge of sports. So, I think you've got to be heavily into sports as a player or spectator. Or I really should say a student of sports. I would recommend a speech major, especially if you want to do play-by-play broadcasting. And you need to get a lot of experience in sportswriting, too.

It's competitive in this area; after all, there are seven or eight reporters here, and only one sportscaster. So be careful before you limit yourself to sports. But if you're good, well, there's nothing better than being right there on the sidelines, spieling out that play-by-play.

Weatherman

At first they had some comedian doing this job. Then they had a beautiful woman. Finally a meteorologist! That's me, Bill Williams, a meteorologist, or weatherman. That's a specialty, and in the last few years more and more stations have turned to the real meteorologist to do the weather forecasts.

If you look around in the studio, you'll see all kinds of scientific equipment and a lot of audiovisual aids I use to plan and broadcast the weather forecasts. I more or less prepared for this kind of job, majoring in science and minoring in broadcast journalism in college. It was a chance for me to combine both interests.

When I first started at a smaller station, I had to double as a news reporter and meteorologist. But I worked hard and finally reached this station where I can concentrate my whole day on weather.

The news reporting, I guess, might come up in the light things I do for the weather forecast. You know, if there is a blizzard, I'll get a cameraman and go out myself to cover the field to see how people are coping, and I use a segment of that in my three minutes of the newscast.

Yeah, you have to have a sense of humor. The comedian is gone from the weather, but you still have to bring a bit of optimism or cheer to the viewers when the weather is really adverse. We even do a few promotional things like offering free snowplows just before winter sets in.

I like science. I like journalism. This is a way I can combine the two.

Other News Jobs

MANAGING EDITOR: This position is also known as assistant news editor or assignment editor. Responsibilities generally focus upon the direct supervision of assigning stories to the news reporting crew. This person works very closely with the news director and implements the philosophy and procedures as set forth by the news director. Qualifications usually require a degree in journalism, communications, or liberal arts. Certainly the position requires reporting, writing, and editing experience as well as managerial skills.

PRODUCTION ASSISTANT: This term might well encompass duties of every position in broadcast journalism, ranging from performing duties as assigned by the news director, producer, or director. It may include a great number of duties, including assisting with the production procedures in tape editing, script preparation, news research, filing, and other newsroom clerical duties.

NEWS CAMERA OPERATOR: Many still photographers aspire to the camera work in broadcast journalism, especially in news coverage. Not only does this require the mastery of photographic and filming techniques, but also a sense of capturing the mood, action, and drama of news events—the emotions of victims in a tragedy, the power and excitement of sports, and the dignity of an inauguration ceremony. As one progresses, he or she is required to have a creative and imaginative eye to film the unusual and the unique angle of perception of what is occurring.

TV ARTIST: Often a news department employs an artist, a cartoonist, or a graphics specialist to provide illustrative materials to be used in a newscast. This usually requires formal training in some area of art and graphics. Often this position is shared by the news and advertising departments of the television station.

EDITORIAL DIRECTOR: Some of the larger stations include this position. The duties include researching, writing, and even delivering editorials on the air. The editorial director might have the responsibility of serving as chairperson of an editorial board that suggests topics and points of view officially expressed by the station ownership.

The National Association of Broadcasters describes other jobs closely related to the news broadcast: The FILM DIRECTOR handles the screening and preparation of all film used at the station and often participates in buying decisions. FILM EDITORS cut, splice, and clean film under the director's supervision. The FLOOR MANAGER directs the performers on the studio floor in accordance with the director's instructions, relaying studio directions and cues. The PROGRAM ASSISTANT coordinates the various parts of the show, assisting the producer–director. This person arranges for props and makeup service, prepares cue cards and scripts, and usually times rehearsals and on-the-air shows.

STUDIO DIRECTORS work on the studio floor arranging sets, backdrops, and lighting and handling the various movable properties that are used in the show. Most stations employ SCENIC DESIGNERS or GRAPHIC ARTISTS who plan set designs, construct scenery, paint backdrops, and handle lettering and artwork. Some larger stations have MAKEUP ARTISTS, COSTUMERS, and other specialized personnel. DRAMATIC ACTORS or ACTRESSES are occasionally used at a local station for a single program or a series of programs.

Most stations have promotion departments headed by a PROMOTION

MANAGER. The duty of this department is to secure publicity for the station, its programs, and its talent. The work typically involves the planning and layout of advertising campaigns and promotional activities aimed at the station's audience. The promotion department may also handle sales promotions, which include the planning and layout of advertising for trade journals and the production of sales brochures and other material used by the sales departments.

The basic job of the CONTINUITY WRITER is to write commercial announcements that will sell the sponsor's products and services, write public service announcements, and occasionally create program material. The first requirement for the job, obviously, is ability to write persuasive copy. A creative imagination is the difference between a fair and an excellent copywriter. A continuity writer must be able to produce material quickly and under pressure. Elaborate rewriting is not always possible.

General Administration

Often newspeople move into the top echelon positions in general administration, and the National Association of Broadcasters briefly describes that area of work:

> This area of a broadcasting station's activities includes the business management and administrative work involved in running the station under the direction of the general manager. The job of GENERAL MANAGER requires a unique combination of business ability and creativity. The general manager usually has had successful experience in sales, programming, or engineering. The responsibilities include the handling of the daily problems of station operations in consultation with the program manager, sales manager, and chief engineer. He or she determines the general policies of the station and supervises the implementation of these policies. Normally he or she handles the station's relations with the Federal Communications Commission and other government bodies and participates in many community activities on behalf of the station.

Network Broadcasting Jobs

The dream of many journalists is the network broadcasting job—with ABC, CBS, NBC. The journalist might aspire to a news jobs or other type of work with the networks. Again, the National Association of Broadcasters briefly outlines the opportunities:

> A network is involved in the same activities as any individual station—producing and distributing programs and selling time to advertis-

ers. Thus, the networks all have the basic jobs found at a local station. Other unique positions are available, however, due to the size of the particular network. Normally network jobs are filled by people with considerable experience. This experience has often been acquired at a local station.

Top network people plan the network's regularly scheduled programing, which is divided into entertainment, sports, and news. The production nucleus of entertainment includes producers, directors, announcers, costumers, scenic designers, and production assistants. Actors and actresses, singers, dancers, and comedians usually are hired on a free-lance basis for a particular show or series of shows. The technical and stage crews, makeup artists, and production coordinators are also part of the production unit.

Network news, public affairs, and sports departments provide extensive coverage of national and international news and special events for their affiliates. Many of the same production jobs available on entertainment programs are available in news and public affairs. In addition, the news, public affairs, and sports departments employ news directors, researchers, and commentators, as well as foreign and domestic correspondents.

Network selling involves contact with large companies and their advertising agencies to plan and execute nationwide sales and campaigns. The sales department is supported by advertising and promotion personnel, as well as the publicity and research departments.

Engineering at the network level requires highly skilled people who are competent in the operation and maintenance of highly technical equipment. A substantial number of network engineers are engaged in experimental and developmental work.

A key ingredient to a network is its affiliated stations. These stations carry programming produced at the network level. Networks have departments which handle their affiliate relations on such matters as agreements to carry programs, clearances for programs, and technical arrangements. All networks own and operate a number of television and radio stations as a division of network activity. Other network departments include research, publicity and promotion, law, program standards, and finance.

Chapter IV

STEVE OLSON—"IT GETS INTO YOUR BLOOD!"

Photo by Greg West

The little guy was the typical little guy in any high school—"too small for football and too slow for basketball."

But Steve Olson wanted to be an athlete, so at Greenfield High School in suburban Greenfield near Milwaukee, Wisconsin, "little guy" Steve Olson went out for wrestling. But something was wrong there too. As he says himself, "I sat on the bench. I could never beat the other guys. I couldn't beat anyone."

Well, there was no desire in Olson to remain on an athletic team strictly for some wild jock identification for himself. He was too energetic, too ambitious. He, like so many high school students, needed success. So, at the beginning of his junior year in high school, Olson began to think about himself—who am I? what do I like? what abilities do I have? what really interests me? what haven't I tried?

Answers to those questions led him to an unexpected conclusion—"I like to write."

In the student guidebook, Olson noted a brief description of the journalism course and decided to enroll. It was a great experience for him, and the summer following his junior year he attended a high school newspaper workshop on the suggestion of his journalism teacher. Returning to school that fall, Olson discovered himself at the school newspaper's first staff meeting. No editors had been selected yet, and the student journalists sat around a table and conducted a brainstorming session for ideas. Olson had the summer workshop experience behind him, and soon everyone was turning to him for ideas. He had them. And it did not take long for the students to realize that their next task was to elect Steve Olson editor in chief for the year.

He reminisces about those high school days, "Working on the newspaper was something special. You soon find out that you're not writing just for a teacher. There are other people reading your writing. It's mass communication. The newspaper hits the hallway of the school and those kids are reading what you say, right or wrong. You get a reaction from students, teachers, administrators. You suddenly realize the responsibility that goes along with working as a reporter or editor on the school newspaper.

"The whole experience makes you feel responsible and important, too. What you write has to be right, factual. It's not the journalism teacher grading it. It's your peers, all the teachers, the administrators. It's a neat feeling to see it in print. You're passing along information, gathered and written by you."

College follows high school—that's an easy decision. But college also demands a major concentration of study, and Steve Olson wasn't quite certain what he wanted. He enrolled at the University of Wisconsin–Eau Claire, a smaller university, which he preferred. He considered a business major, but he explains, "I saw some of the people with business majors and they didn't impress me. Students were actually discouraged from majoring in business, and the professors made their courses tough to dissuade students."

At first Olson explored microbiology, but soon his high school memories and experience in journalism led him to switch to a journalism major. He laughs at himself as he says, "One of the first things I had to do was learn to type."

Olson started in news editorial journalism to prepare for a newspaper career, but he soon became disenchanted with the program. He explains, "The professors were too concerned with style in copy. They never really dug into how to report the news—that you should take a national story and try to localize it. I don't think they steered you in the right way. They were too concerned about style."

Steve wanted more writing courses. One day his college roommate suggested, "Why don't you change from the print media to broadcast journalism? You have the looks to be an anchorman."

Olson says he told his roommate, "Hey, I want to be a writer, not something on television."

But he did switch.

And that was really the beginning of college for Steve Olson. He gives much credit to Henry Lippold, a broadcasting professor who is described by Olson—"He's a dynamic person who sleeps only two hours a day. His sheer energy reached into you. He pushed you. He'd call you up in the middle of the night on a breaking story. For him you had to get out of bed and cover the story."

The university sponsored a news program entitled "Update" twice a week on a local cable television station. The broadcast journalism students reported, wrote, edited, and broadcast the news for that program. Students from speech classes performed all the technical work, so it was totally a student production. This was much different for Olson from the news editorial program, as he says, "You'd make a mistake and they'd tell you what you did wrong and how to correct it. We'd watch network television news and critique it."

While at the university, Steve Olson also did radio work. "I did sports at first. That was fun because I have always enjoyed sports. But I reached a point where I had to decide whether I wanted to go into sports or news. I like news better because news can be both hard and soft. There are so many more opportunities in news than in sports."

In 1978 Olson obtained his first internship on the WEAQ radio station in Eau Claire, where he soon adopted some attitudes that would lead him to a good job later. He recalls, "I'd do stories no one else would do. The news director would give me feature material, and I'd always look at it as a back way to make a hard news story. I'd take a national issue such as pork prices and get hold of some state officials."

Olson's start in broadcasting was typical of many other graduates—several short stints on small market stations. After several jobs in Eau Claire and LaCrosse, Olson finally in 1981 moved up to the medium markets with a general assignment reporter's position with WHBF-TV, a CBS-TV affiliate in Rock Island, Illinois, where he also performs radio news anchoring chores on WHBF radio.

The general assignment reporter may have any topic assigned to him, and that can be tough in the task of interviewing both officials and people on the street. As Olson looks back upon his education, he believes he would have included more business and economics courses. He says, "If you go into just journalism as I did, you'll be missing so much by taking only what is required. You either know how to write or you don't. Out in the real world it's all practice. It's more or less what you have stored in your brain or what you've learned somewhere along the line that helps you in an interview situation. You interview a school board president one night, and the next night you have some farm official or scientist on corn crops. You've got to know all these bits and pieces—just enough to start that question and get the interviewee to explain it. And even if you don't know what you're talking about, you've been exposed to it along the line."

The reporter needs to be very conscious of ethics, Olson believes, noting that he wishes he had had courses in press law and journalism ethics. He explains, "Let's face it, you're dealing with some pretty powerful stuff as a news reporter. The cliché that the pen is mightier than the sword is true. Look at the politicians who have been booted out of office because of news coverage. I think I should have studied economics more. Back in college, I didn't think it was that important. I ignored business courses as if they were the plague because everyone else was majoring in business. I didn't take any political science courses until my senior year, and now I wish I had taken more. I'd want more courses that would make me think."

Those interviews come fast for Olson on his tour of reportorial duties because he is on the night staff, which many times means he must do a story for both the 6 and 10 p.m. newscasts on WHBF-TV. "Those stories have got to keep people's interest on television," Olson says, "In a story I want it visual enough to carry a TV story. There are good stories out there that don't make good television picture stories to carry through from a beginning to an end."

That is one reason Olson loves his weekend assignment on WHBF radio, for as he says, "You don't need visuals to kep the people's attention in radio." He looks upon television as "too showy," quipping, "It's entertainment with a sprinkling of news."

The show-biz aspect of television causes Olson both a bit of amusing reminiscence and stark reality. He says, "I'm conscious of the cosmetic aspect, and it bothers me. I used to have long hair. In fact, once I had to tie it in a pony tail. It was one summer in college when I was working at a truck terminal, and I had to use a rubber band to tie my hair under a hard hat. I'm very comfortable in blue jeans, but here you find yourself having to dress up in a three-piece suit, and things like that feel confining for me. Oh, I don't mind it so much now. As you start wearing suits, you become used to it."

More important, the show business influence bothers Olson as a reporter, and he explains, "I'm not here to be show biz, but it's here and you must be saleable to an audience."

Being only twenty-five years old would seem to be a cosmetic benefit, but if it is, it is soon offset as a handicap as a reporter. Olson says, "The thing that is difficult for me is that I look so young and that influences my credibility—you know, was this kid born yesterday? They tell me I should try anchoring, but right now I'd rather be a reporter. That's the bread and butter of every newsgathering operation. That's fun! You're out in the field, gathering those stories. That's where the action is."

But there is more than the cosmetics, and that is the attitude of the reporters in their approaches to their work. One of those attitudes certainly must involve the social consciousness of the reporter. It is evident that Steve Olson demonstrates this attitude as he expresses his desire for a certain type of feature story, "I like the kind of story the average reporter says can't be done. The story on how to do something. I like that story that can move people to a social change."

The kind of story that is the most difficult for Olson is the follow-up on a tragedy. He recalls a furnace explosion in which both adults and girls attending a 4-H meeting were seriously burned. He says, "I was there to cover the first story, which was sold to network television. I followed up by visiting a father who had suffered extremely serious burns.

There he was, in good spirits. But I stood there and feared he saw me as a ghoul chasing a tragic story."

Another story that riles Olson is the follow-up to a story that has appeared in a local newspaper. He explains, "The assignment editor will ask me to do a story with the same angle as was used in the newspaper. I really resent that. So I go out on it and there's nothing there. So I call him back at the newsroom and he tells me to do something! But sometimes there just is no story, and you have to tell him so. I don't like to see us do a story just because the newspaper did it the day before. I want to get out and get the story and have the newspaper follow us."

But Steve Olson will do any assignment handed to him, and he likes "getting out into the field." He praises his news director, saying, "He has the philosophy to stay away from the officials. We can contact them on the telephone. He wants us to get out and talk to the people. He says that we can call the president of John Deere about a layoff and he'll tell us the company is doing fine. But we need to go out and talk to the workers who were laid off."

It takes a competitive and aggressive person to be a successful television news reporter, and Steve Olson recognized that as soon as he moved to WHBF-TV in Rock Island, which is one of four major Illinois and Iowa cities that comprise the Quad Cities metropolitan area.

The young, handsome reporter smiles as he recalls his first experiences. "When I started here, I was intimidated by the two other television stations. This was the first time I had real competition. Here I saw older reporters. There were TV logos all over. Reporters were dressed up. Every reporter had a cameraman. There were station news cars. There were reporters from three competing daily newspapers in the Quad Cities. It was all very intimidating. But I wasn't going to let them know I was intimidated and overwhelmed. I was going to go out and cling and claw with the best of them even though I wasn't familiar with most of the officials and ongoing issues. I listened to the questions other reporters asked and tried to learn fast."

Learn fast he did. Olson quickly mastered the skills of backgrounding the news—a responsibility for any reporting assignment. And he enjoys that backgrounding as much as the actual interviews. He says, "If I can't go far on television, I'd like to go to the networks and do research. I'd like to do the investigating behind the scenes. Mike Wallace on 'Sixty Minutes' is just the frosting on the cake. Others do the digging, and I could enjoy that. I think I have imagination, and you need to think obscurely at times. My imagination helps me, I think. There might be a wire story on fallout shelters and I try to imagine what I can do on that locally."

Then there is loyalty and dedication to a television station, which is a must for the reporter. Olson demonstrates it as he says, "Our station slipped a bit in the last Nielsen ratings. I was really disappointed. I had worked overtime—worked hard! I want the viewers to watch our channel. I want them to know we're a good news station. I wouldn't want to work for a number three rated station."

Such an attitude requires a television news reporter to make constant self-assessments, and Olson does not shy away from or avoid his weaknesses. He knows he wants to move up to the major markets and possibly to network television news.

As he looks at himself, he says, "I've got to learn to get good enough so that I can think of my shots in the field. I must also reach the point where I don't put the burden on myself. I still feel that I've got to edit my own story, to cut the voice track, to put it all together. In the major markets the reporter cuts his tape and that's the last he sees of it. I've got to learn to trust someone else with my story."

Reflecting momentarily, Olson continued, "Yes, I've seen failure. I need to improve my standup. I need to be willing to work with other people. You can't come into a news operation with the attitude that you know everything. You'll upset people. You should ask a lot of questions; that's how you learn how things are done and why they are done that way. You've got to be a team player and unselfish."

The future? Well, like most other television newspeople, Olson wants "to work up to the biggest television markets."

Will he stay in it? Probably. Especially as he finally says, "I love it. It seems like a hobby I happen to get paid for. It gets into your blood, the writing, the deadlines. I always hope stories will move people into some kind of action—like seeing people go to the aid of someone in distress."

Pausing, Steve Olson says, "Yeah, I love it. It's neat seeing people act."

Chapter V

CAROL CLARK—"MY FIRST STORY—IT WAS TERRIBLE!"

Photo by Rod Vahl

She was straight out of college.

She had just completed a four-month internship.

She was only twenty-three years old.

She was a woman.

She was a black.

Now, in television broadcasting, all of those forces are powerful hurdles to cross in a world that is dominated by the experienced, seasoned, white male journalist.

But the opportunities are there, and Carol Clark knew but one dream— to be a television newswoman. From her elementary school days, Clark would always respond, "I want to be a news reporter" when asked what she hoped for her future.

Carol is attractive, intelligent, ambitious, determined, and patient.

All of those attributes are essential for the student with dreams of invading television, which offers chances for worldwide travel, excitement-filled assignments, and nationwide recognition. Carol Clark harbors such dreams, but she also is fully aware that reporters are hired and fired as fast as the national and local audience ratings are published. "I'm constantly conscious of that point," Clark admits. "I've been here for two years, and already I've seen a good number of reporters come and go."

"Here" is WQAD-TV, an ABC-TV network affiliate in Moline, Illinois, a part of the Quad Cities metropolitan area that encompasses medium-sized cities in both Illinois and Iowa. The TV station is a medium-sized market, and Clark acknowledges that she was very lucky in bypassing the three to five years that many television reporters must work in the small market television stations.

But it was not all that easy for the hard-working journalist. In high school in Rock Island, Illinois, there were the disadvantages she feels so many blacks suffer because of what she terms "the white teachers' failure to understand and to help black students." But she was "lucky," she says, "because I ran around with kids who were headed in the same direction I was."

Clark reflects upon her childhood days and her years in high school. "I was raised by my grandmother, and when I entered junior high school I discovered I liked to write. I took all the writing courses I could. It seemed writing was the most interesting thing to me and it came relatively easily. So I just started saying I wanted to be a writer."

But friends cautioned her, "You won't make money as a writer. You'll have to struggle. You'll have to crawl before you can walk."

Clark remembers saying to herself, "I'll do it. I'll go to college and study. I'll take every course I need." And she did. Upon graduation from high school in 1975, Clark enrolled at Illinois State University in Normal and majored in mass communications.

But once in college, Carol Clark ran into an important barrier. No, not racial discrimination, but another social reality. As she says, "It seemed to be whom you knew, not what you could do."

Clark referred specifically to the opportunities to work on the school's cable TV station. She reflects, "It was whom you knew and if you were in the right place at the same time. If not, you did your stuff for class purposes only. I did my work. I just didn't have a good relationship with the instructors in the broadcasting department."

That bothered the co-ed immensely. She says, "There are students who want to get ahead and be somebody. They will put in that extra effort. We all want to get ahead, but a lot of times you find students who will do anything to get that spot. But there are people who just don't sell out themselves. I wanted to do things on the college television station, but there were always those who got the spot before me."

It remained that way for Carol Clark throughout four years until she was graduated in 1979. She returned home and went to all three television stations in the metropolitan area, but she found no good fortune. As she explains, "I talked to the news directors and told them I was looking for a job, that I had just graduated, and that I'd do anything to get my foot in the door in broadcasting. But they all told me the same thing—they didn't have any openings. They told me to try the smaller TV markets, that one just doesn't start in the medium-sized markets."

The summer was soon gone, and Clark still had her foot outside the door. But a stroke of luck emerged. She met a black engineer who worked for WQAD-TV, and upon learning of her bad luck in applying for a job at all three of the stations, he asked her, "Would you be willing to take an internship to get that foot in the door?"

There was no hesitation for Carol Clark in responding in the affirmative, and two weeks later, October 22, 1979, she was offered an internship at WQAD-TV. "I couldn't sleep the night before." she says. "I was too excited. I knew I would now have a chance."

That chance offered Clark a world of experience and some moments of frustration. She ran errands, filed scripts, went out to observe reporters. One month passed into two, three, and four months. The frustration crept within Carol, who constantly asked herself, "Will it come? Will I really have a chance to be a reporter? Do they like me? Do they think I can do it?"

At the end of the fourth month, in February, 1980, all those questions were answered for the aspiring reporter by an offer of a full-time job as a general assignment reporter.

Filled with ambition, Clark pursued her duties with zest, but she soon learned that she had much to learn. She recalls her first TV news package assignment. "It was a story on a paramedic program that the city of Moline was trying to initiate. It was a night meeting of a lot of city officials. I covered the meeting, wrote the story, and voiced it. It was terrible, oh, so terrible!"

That meant that if Carol Clark were to survive in television news she needed to take a hard, close look at herself.

She did.

"First of all, I saw I had a problem with my voice," she explains. "I have a soft voice, and I knew I had to build it up strong enough to make people listen to me. You need to develop a style for yourself in TV broadcasting, and it wasn't there. No style!"

To eliminate that weakness, Clark bought a tape recorder, and every night at home she practiced with her voice. The next day she would take the tape to work and ask anyone and everyone to listen to her voice and to critique it. She says, "They were great! They'd tell me if something wasn't right. I would underline words that needed to be projected more. I really worked at it."

Then there was the writing. She was so dissatisfied with the very thing she thought came easy to her. She explains, "I discovered I really didn't have a style in my writing. It was basically just words. After I wrote a piece I would sit back and analyze it myself. I tried to pick out words and phrases that I could change. I'd look at a story I did the day before. I'd ask myself if I could have changed something to make it easier for the viewer to understand. How could I have made this story relate more to the viewers?"

Such self-evaluation requires the aspiring journalist both to be self-critical and to be able to accept criticism from others. Clark explains, "People don't put you down, but if something isn't right with your story, the next day when you come to work you'll hear about it. And it's not in a way to make you feel bad, but in a way for you to better yourself."

She adds, "I found out when I first started that if you aren't willing to take criticism, you're in the wrong business. There are some weeks when every story you do has something wrong with it, and somebody will point it out. If you can't take it, you can't take the business."

Though she has much confidence in her writing ability, the young reporter seems to concentrate much of her attention upon that aspect of her work. She says, "I really need to develop my writing more. I need to be more creative. We have one minute and thirty seconds to tell a story, and it's tough to limit yourself to that time, but that's part of the growing in your style. You learn. I can tell every time I've improved my technique. If something goes right for me and it shows in my story, I know it. And then I can say to myself that I can go on. I listen to my delivery. I look to see if my story is flowing. You can tell if the smoothness is there."

Looking at the competitive aspect of television broadcasting, Clark admits, "I thought there would be competition in the newsroom. During my internship it seemed that people wre competing, trying to be the top reporter. But that was looking in from the outside. When I became a reporter, I didn't see the competition."

But it was very different in the competition with other television stations, as she explains, "When we're out on a story with other TV stations, we do compete. I look to see what other station reporters did. I remember when I was covering a fire I found an eyewitness first, and when I saw other reporters coming up, I was trying to hide my key eyewitness from them. But it didn't work. They came right up. We do try to find the exclusive report. We have to develop our sources. That's hard when you start out, because you don't know people. But you soon meet the officials and the power-structure people. You develop your sources, good sources, and in reality, most are willing to help you."

As with all in the television news business, the show business factor arises, and it has been no different for Carol Clark, a beauty in her own right. She says, "At first I didn't think about it. I put it aside. I had so much to work for. I still put it aside, because I just need to be a good reporter. I don't think about being the TV personality. I'm more conscious of it now, because more people are beginning to call me and I get Christmas cards. People say they watch me. They see me more now and they are aware of who I am. I'm getting more attention now."

She quickly casts the show-biz bit aside, saying, "I have to pay my dues first before anything big happens to me. I started in a medium-sized market and that's tough. Oh sure, I know what I want to do—to be a correspondent for a network. At first I thought about being an anchorwoman, but now I want to report, be out in the field and travel. That would be exciting. I like the spot news—being out there where the news is happening. I still need to look for more opportunities to expand upon my writing. As a general assignment reporter we get to cover a lot of things. I wouldn't want to have a specialty; I think that would restrict me. And I want the on-the-spot news so I can travel. I haven't seen much of our country, and I want to see it. I think I know what's going on. I read a lot, and I watch all that's going on. And I want to be right there, reporting it."

Carol Clark has never felt that she has experienced blatant racial discrimination in her work, but she does say, "I think it's harder for blacks in television. I think we have to work 110 percent where some whites work only 90 percent. I had to think twice at first if I wanted to go into this, because I had to go through so much to become a reporter. It crossed my mind many times whether others had to go through this or if I was going through it because I'm black. Well, I still haven't figured that out, but it hasn't stopped me."

Recognizing that many black students wonder if there really is a chance for them in television news, Clark offers advice, "First of all, no matter what you do, you've got to have a positive attitude. That's the kind of attitude I had. I knew when I started that internship I was so excited that I was on a cloud for two months just because I was in the newsroom."

"You have to have the right attitude no matter what happens at the station or where you are working. If you can bear with it and go on and say the next day will be better for you, you've got it. That's what I had to do. It's not because I'm black; it's simply because I'm new and I'm learning. I think that's what got me through. As I kept working during that internship, I felt like just an errand person. I wasn't getting anyplace. I kept saying to myself that one day I'd look back at this and ask myself how I got through it all."

After a pause, Carol adds, "As Martin Luther King said, 'We shall overcome.' I feel that I have. I really do. I can look back now and say I really have."

The young reporter feels the black can play an important role in the news coverage of black issues, but she cautions, "When I first started to cover stories, I felt I was being given all the black stories, and I really resented that. It's one thing to be used as a reporter and another to be exploited. I think sometimes on a black issue you may need a black person to cover the story. Sometimes we use a jargon that a white reporter might not understand. There are instances when you do need a reporter who can relate to a certain people. But there is a difference between those stories and sending me out because I'm black."

But Carol Clark is quick to point out that she does not feel she is being used to fill any kind of a quota. The station for which she works has more minority-group reporters than either of the two other television stations in the area. She says, "But if you want something bad enough and you are there simply to fill a quota, stay there! I would, simply because I could accomplish what I wanted to accomplish. We'd both get something out of the deal."

The very fact that Carol Clark does not dwell upon the problems a black might encounter in seeking a television broadcasting career vividly demonstrates her fast-growing reporting skills.

She is aggressive: "I can't sit back. If you wait for people to come to you, you won't get the story."

She is patient: "I know what I want to do, and I can wait until I learn all there is to know to get where I want to go."

She is curious: "I am inquisitive and I want to know the why of every question that comes up."

She is self-confident: "I had to work on this. But I soon learned that when you have it, you'll feel it. It clicks inside of you."

Finally Carol Clark says, "I love what I'm doing, and I don't think I could do anything else. I never want to do anything else. It's exciting, and something different happens every day."

Chapter VI

ADLAI STEVENSON—"A REPORTER'S MOST VALUABLE ASSET IS HIS COMMON SENSE."

Photo by Rod Vahl

Adlai Ewing Stevenson IV.

First, delete both Ewing and IV.

Second, quickly recall and then forget that this is the son of Adlai Stevenson III, former U.S. Senator from Illinois; grandson of Adlai Stevenson II, former Illinois governor and two-time Democratic candidate for President of the United States; and great-great grandson of Adlai Stevenson, Vice President of the United States under Grover Cleveland.

Yes, that is an ancestry solidly etched with political gold all the way from the Chicago precincts to the Oval Office in the White House. But for Adlai E. Stevenson IV, the modest weekly paychecks earned from pounding the streets for news stories are far more to his liking than seeking the silver- and gold-emblazoned emblems of political office. At twenty-six years of age and with a self-acknowledged "mediocre" college career, young Stevenson haunts the city hall chambers and corporate executive offices as a general assignment reporter for WMBD-TV, a CBS affiliate station in Peoria, Illinois.

A broadcast journalism career was not a lifelong dream, fantasy, or ambition for Stevenson. Rather, one suspects the Stevenson youth was easily dubbed one of those "rich kid" stereotypes who did a good deal of playing and very little studying.

Stevenson quickly admits that his academic life was not of great achievement. "After grammar school in Chicago, I enrolled in a college prep school—Middlesex in Concord, Massachusetts. I didn't have any special interests, and it wasn't until I enrolled in Northwestern University that I discovered a special interest of any kind."

It was the camera that suddenly captured Stevenson's attention and labors. He says, "My roommate was an editor on a Northwestern publication, and I became the photo editor. I must admit I got the position not so much because I was a great photographer as because the editor was a friend of mine and I just became heavily involved with photography."

Stevenson's interest in photography led him to drop out of Northwestern for six months to work as a staff photographer for the *Pantagraph* in Bloomington, Illinois. Again Stevenson emphasizes his "connections," saying, "I should make it clear that my family was part owner of the newspaper, and I got the job partly because of connections."

After the six-month stint with the *Pantagraph,* Stevenson returned to Northwestern, from which he graduated in 1978 with a degree in communications studies. But like so many other graduates, Stevenson had no real idea what he should do with his life, so for two years he worked in Chicago bars and restaurants while he pursued a free-lance writing career.

Those efforts were mostly nonproductive. He explains, "I managed to see a few articles published in the two Chicago newspapers—the *Sun-Times* and *Tribune*—but I found free-lancing very frustrating. I wasn't having much success. It was especially frustrating for someone who wasn't very good at it, and believe me, I wasn't!"

Again, as many college graduates will do when having no genuine direction for themselves, Stevenson decided to return to the classroom. For him, it was back to Northwestern to earn a master's degree in journalism. Probably for the first time in his life, Adlai Stevenson really worked, because he both studied at Northwestern and worked as a reporter for the Chicago City News Bureau, which feeds stories to both the *Sun-Times* and *Tribune*. Stevenson describes that year from 1980 to 1981: "I was quite proud and happy working for the City News Bureau. It's a famous starting point. A lot of famous people got their start there. You learn a lot in a practical sense. You just pound out the stories every night, and I loved every minute of it. I spent the evenings in police stations and in morgues. In hotel suites with Jimmy Carter and Ronald Reagan. It was great, but it wasn't easy."

Stevenson does not reflect upon his college training with great enthusiasm, but rather as an essential requirement "to get a job." He explains, "I thought half of my studies for a master's degree were an immense waste of time. I had to take introductory courses I didn't want, but Northwestern had its rules. When I got into the technical part of television, I thought it was good. The theoretical stuff didn't interest me to the extent I had to take it. I liked the vocational training, but I did not like the theoretical and abstract. Every college has it! At least 75 percent is not really worthwhile. I don't want to attack the college. One has to remember that college is best for getting you a job. It's unfortunate you can't get a television job without the degree. I knew City News Bureau people who were very talented but had no degree. That's why I went to college—just to get a job!"

Despite his seemingly strong resentment of academia, Adlai Stevenson's words about himself somewhat belie that resentment, as he says,

"I'm an idealist about my practicality. I'm not a liberal. I'm not a social reformer or a conservative. My value judgments are based simply upon the idea that I must keep my common sense and use my head. Where is the story? What is the story? Where does the story lie? I admire Bill Moyers—if he is an advocate, he's an advocate of ideas. Ideas will speak for themselves if you lay them out clearly. That's a big challenge in itself. I feel I've got to wade through all the crap and get to the heart of the matter."

But why television news rather than the print media for Adlai Stevenson? "I'm more suited for television," he replies. "I have a good voice. But more important, it was the difference in the job markets. The newspapers just don't have more jobs. Television is more volatile. Take cable television—it will shake the iron hold the networks now have. Let's face it, I want to go with the job opportunities. I want to stay five to ten years in the news business. You've had enough unless you're among the chosen few who are nationally known. I'd love to become a network reporter. That's my goal right now. I'm not as flashy as other local reporters. The networks are a bit more subdued and serious. That's more my style."

And it seems to be true.

Adlai Stevenson is a handsome young man despite the balding dome he inherited from his famous father and grandfathers. He is a spitting image of his father, with deeply set dark eyes that seem constantly in anticipation of discovery and enlightenment of one's own self-realization.

And that Stevenson "practicality" is evident in his approach to his work as a broadcast newsman. In Peoria, Stevenson's regular newsbeat is city hall and several giant corporations that are much of the source for the Peoria economy. The reporter describes his job, "I do a lot of labor layoff stories now. I'm limited in time, but you'd be surprised what you can say in 60 seconds. You can tell the story despite the time limitation. A well-written story can have an incredible amount of information. No, not a lot of facts and statistics, but you can tell the heart of the story. In a layoff story I'll use a closeup and personal look at one family. I like to tell the economic effects of the layoff. Just what a layoff means to the daily lives of the people."

Stevenson describes the importance of being practical in one's approach to the television news story: "In TV you narrow the focus and keep it all in a straight line. You have to know how the story is going to come out the minute you go to work on it or the minute you go out on the street to gather your material. You must conceive the story in your own mind and direct all your energies toward your original idea for the story, because you don't have time to come back and sit around thinking what you're going to do with it. You need to think beforehand. To a certain extent that might be prewriting the news, but in those spot-breaking news events, you have to think quickly."

Such demands are why Stevenson rejects the theoretical for the practical, and he says, "One needs a general wealth of knowledge. But a reporter's most valuable asset is his common sense. You can get so tied up in procedures and theory that you lose track of what you want to do—which is, to tell a story!"

Whether or not the television news broadcasters "tell the story" as adequately as newspapers is a continuous debate in journalism circles, and Stevenson has his views: "I don't think it's a productive argument. They are different. They both have their strengths and weaknesses. When I was in the print medium at the City News Bureau, we would snub the television newsmen and laugh at them. Now after being in television news and seeing the amount of work that goes into a news report, I'm very impressed. You realize the production aspect. I think we do a good job in getting out the news when one considers the constraints of time. People miss the point when they criticize television news in relation to print. The biggest criticism against both areas is not how they cover the news, but what they cover. There is a whole issue of gatekeeping that is little touched upon."

Stevenson believes strongly in television news and says, "Most people get their news from television. The problem is that television is often only a fleeting impression; you have to make sure you have a simple, strong message, because if you don't, the story may be lost on the people who hear you. What you lose in content and detail, you make up in impact. You tell it in a different way. In television, you give it a narrative line. You tell what's happening without much background. I think most people get as much out of TV news as they do from print."

Indeed, the reporter recognizes the weaknesses in television news. "I'm sure bad TV gives the emotional impact and not facts. But there are many serious stations. When I go to my next job, I'm going to look carefully and go to a station where I won't have to do embarrassing things and where I won't have to do stories about incest, rape, and so on."

Stevenson despises the sensational news stories; he says, "I'm frustrated by this, which I call bad show biz!" Much of the show-business aspect arises during ratings time. Stevenson recalls, "In Chicago during one ratings period, all the television stations focused upon teenagers and sex, teen prostitution, and one station even on sex in rock 'n' roll music. Maybe all of those stories needed to be covered, but all three stations went after teenagers and sex at the same time to prove they can do a good job."

But Stevenson doesn't agree with some critics that ratings such as the Nielsens are bad. "I recognize it's competitive at the network level. Let's say that show biz does not necessarily exclude good journalism. CBS-TV is one of the better examples; they have the best news reporters out there getting the stories. Take Dan Rather; how often does he make a mistake? Sure, they have him put on a sweater and prop him with some graphics, but those don't conflict with the good news stories they present. There is no conflict between good news and a good show."

One demand on the broadcast journalist is that constant self-evaluation. Whereas the print journalist need not give great concern to such factors as speech delivery and cosmetics, the broadcast journalist must. All recognize the show business influence, and Stevenson is no exception. "It's difficult!" he says. "No matter what the story is, I never like it because I always see things that could have been done better!"

What he dislikes most is seeing himself on television. "I've got to get used to seeing myself if I'm to stay in this business. I try to evaluate myself in a cold, harsh manner. This becomes a very personal business. People judge you right away. But I'm concentrating on my voice. You can't change what you look like, but you can change what you sound like. The one quality that is most important is the voice. If you talk well, the audience will listen. They don't care what you look like, but you do have to be articulate and concise. I know I need to relax more in front of the camera. Oh, I'm not worried about that, because I know it will happen. I feel more confident now."

In addition to that good voice, Stevenson advises young people seeking a broadcast journalism career to:

—"be pleasant, believable, honest."
—"be spontaneous without flubbing.
—"think quickly on your feet."
—"maintain a reasonable appearance."
—"write in a simple, narrative style."
—"think and plan in a logical order."
—"read a lot to gain a great depth of knowledge and awareness of what is going on in the world."
—"be practical, not theoretical."

Then, Stevenson adds, "Trust your own judgment. There is a pack mentality out there, and often the pack is wrong. You are going to make it or fail on what you do in the long run, so don't make what you do following around in other's footsteps. You've got to think on your own feet and ask what you think are the right questions."

Adlai Ewing Stevenson IV.

"My name? Yes, I run into it constantly. Yes, I can get a job because of my name. But I have to do the same things others do. I have a job to do. In the long run, people will rate you on what you do, not your name. I work around that pretty well."

Indeed he does!

Chapter VII

JANE PAULEY—"IN AN INTERVIEW, I'M THE LEAST IMPORTANT."

Photo by Joe Glowacki

"FIVE, FOUR, THREE, TWO, ONE. You're on!"

Jane Pauley tosses aside a section of the New York *Times*, grabs a pencil, smiles into the television camera, and says, "This is 'NBC Today,' March 14. Tom Brokaw is in Florida covering the primary elections. Gene Shalit will be visiting with Marsha Mason, star of 'Chapter Two,' and we will be talking with William Whyte, author of . . ."

Ah, the world of network television looks so glamorous! Especially with "NBC Today" correspondent Jane Pauley sitting before the TV cameras, seemingly talking so personally to each of the millions of viewers about the breaking news of the day, the latest episodes in the lives of celebrities, and the newest novels to hit the best-seller lists.

And indeed it is glamorous—both for the viewer and for the TV personality. But for the TV journalist, it is much more than sheer glamour. There is much hard work and spontaneous action in newscasting, as Jane Pauley explains, "You've got to get off the mark fast. It takes time to warm up a guest, to mine the best material as it were, and we don't have the luxury of mining. It's 'go get something and get it out fast.' Most of the time it works, and sometimes it plainly doesn't. It requires many disciplines."

Those disciplines in preparing and conducting live interviews have not come easily for Pauley. After an outstanding high school career of extemporaneous speaking in the National Forensics League in Indianapolis, she earned a degree in political science at Indiana University with the intention of seeking either a law degree or a television broadcasting career.

Upon graduation, Pauley became involved in Hoosier state politics for about eight months before joining WISH-TV in Indianapolis as a reporter. Soon she was co-anchoring the midday news reports and anchoring the weekend news telecasts.

At twenty-four years of age, the young journalist was selected to be the first woman to co-anchor a nightly news telecast on NBC's WMAQ-TV in Chicago, a position she held until NBC signed her to a contract to be the regular correspondent on "Today" with Tom Brokaw—at the age of only twenty-five. This year, she was named principal writer and reporter for the Saturday edition of the NBC Nightly News.

Pauley is proud of her speaking experience in high school, referring to that era as an "aberration—when I wanted to collect blue ribbons, medals, and trophies." But, as she remembers, "The most important thing I did in high school and college was my extemporaneous speaking.

"On one hand, I'd get up early in the morning and get on a bus and go to Fort Wayne or up to Ball State University in Muncie and get off the bus at a speech meet, feeling cocky because people would look at me and say, 'There's Jane Pauley!' By the time I finished high school, I was one of the best in the state."

But there was another side, as Pauley continues, "On the other hand, my stomach would be so upset, I'd have to go to the ladies' room and be almost sick in fear. So it was that mixture of 'hotshot Jane' and the utter panic and insecurity that made me sharp enough to do it."

Pauley has advice for the student seeking a journalism career: "If I had it to do over again, I would double major something with journalism, not just major in political science. That's a foundation course for something else, such as law school. I would recommend a double major—journalism with political science or with English."

The blonde newscaster really felt the lack of journalism training when she suddenly discovered herself upon a fast track in her career, being snapped up by NBC for national television while she was barely getting her feet wet in Chicago.

She says, "It was like skipping grades or graduate courses. I couldn't give NBC excuses about being so young. They expected performance. I just wish I knew more about production. If I had been a reporter in the field or a correspondent on the road for a few years, I would know more about how to produce a good news story. I still have a fairly good sense about it. And I'm a fairly good editor of other people's copy, but I'm aware that I do have some shortcomings because I was on such a fast track."

Despite those shortcomings, Pauley kept her nose to the grindstone at all times, constantly seeking to "catch up" on those demanding journalistic skills. Above all, she quickly mastered the attitude that a good reporter is always on duty, no matter if she is on an airplane, at a movie, or dining out with friends.

After the "Today" program each day, Pauley heads straight for her office at NBC and commences what is almost a daily ritual. "After the show I normally spend part of the morning reading the newspapers—the New York *Times*, the Washington *Post*, and the New York *Daily News* for the TV page. I flip through the *Christian Science Monitor* every day, because they can summarize an international event onto one page about as well as any newspaper I know."

With the change in the economy, Pauley now spends many afternoons with the *Wall Street Journal*, explaining, "The economic and financial news suddenly is more important than it used to be, whether it's inflation or the connection between economics, the price of oil, the Iranian revolution, and the value of the Persian Gulf."

Whatever it is, Pauley nurtures that attitude of always "being on duty." She says, "I'm always looking at something and wondering how it can be translated into our 'Today' format."

Another reporter's skill that Pauley recognized she would need to master was journalistic objectivity, and it wasn't easy for her. She attempts to harbor no pet causes, feeling that not having causes is an advantage in her work as a news correspondent.

She explains, "Like anybody, I have my politics, my personal political orientation, and I have opinions on a lot of things, but I don't belong to any women's organization. Although I feel rather strongly about the women's movement, I would probably call myself a feminist with a small 'f'. I don't organize. I don't think it's my place to stand on the platform for ERA, for instance. It's my ability to stand apart from causes and movements that I think allows me to look at them with the objectivity a journalist needs."

Usually that objectivity requires Pauley to literally step aside in the interview and allow the visitor to have the spotlight. She explains, "The Johnny Carson interviews are different in that he is the entertainer. He never does an interview in which he is not the most important part. But in my interviews I am the least important part, and for the most part, the viewers should not be aware of me at all."

Pauley's least favorite interview, which would probably surprise most viewers, is the celebrity interview. She explains, "They are the hardest to do. It's hard to say that except for your new movie, why are you here?"

She can point to exceptions such as a recent interview with Kristy McNichol of the TV "Family" show. Pauley says, "I thought my interview with her was delightful. I was pleasantly surprised that she was as articulate and grown up as she was. I was very pleased that she had something to say, that it wasn't just 'What it's like to be a movie star,' but that she wanted to talk about why she felt her new movie shouldn't be R-rated. She made it easy for me."

But Pauley's favorite kind of interview is what she would call a "pressure performance." Because Tom Brokaw was in Florida covering the primary elections in March, she went to work in the New York studio at 5 a.m. to write the first newscast.

When she arrived at work, she received a surprise: "I learned for the first time that morning that NBC had a satellite connection to Teheran and that I would be interviewing Iran's foreign minister, Sadegh Ghotbzadeh. I couldn't do any research. The networks had been back in Teheran for only a couple of days, and their situation was in such a state of flux that there wasn't a set of obvious questions.

"This was the first live interview with Ghotbzadeh in weeks, and I was doing it! It worked out well. I suppose it could have been a disaster, but because it worked out well, I was satisfied that I had asked the right questions. It was such a pleasure once it was over. The feeling of satisfaction you get from that kind of interview is more than you get from any other kind."

The future for Jane Pauley? Well, she says she is not a goal-oriented person. Rather, she explains, "I see my career more as a process than a series of goals. I certainly never said I wanted to do the "Today" program ten years ago. When I was in college I would have said I wanted to do the news on WISH-TV in Indianapolis, and though I knew I would leave there sooner or later, I never anticipated to have what I do today."

Jane Pauley perhaps best summarizes her whole attitude in saying, "I see my future not so much as what I will do next as growing into that role where I simply say, 'This is Jane Pauley, NBC News, New York.' "

Chapter VIII

PETER STURTEVANT—"I LIKE TO KNOW THINGS FIRST."

Photo Courtesy of CBS-TV

There is an excitement merely in the way Peter Mann Sturtevant, Jr., sits at his desk, constantly scanning the leads of wire copy stories that pile higher as the day progresses. His passion for the news is obvious as he conducts telephone conferences with instinctive, terse responses and orders to the eight CBS-TV news bureau chiefs whom he commands from the New York newsroom.

He listens with eager anticipation to a breaking story, he observes with an analytical sense of what is news, and he selects with an urgency to inform the people of that news as quickly as possible.

As he stands to point at a U.S. map, he is a tall, athletic figure who could easily be taken for a powerful New York Yankees pitcher or a volatile opponent of Björn Borg on the tennis courts—both fantasies to which he often jokingly refers in casual conversation.

One wonders why this handsome, bearded thirty-eight-year-old newsman is not on the TV tubes broadcasting the news rather than compiling stories for the evening newscast. But that is not his game; he simply wants to be in that news spot where he learns firsthand what is happening.

Peter Sturtevant is national news editor for CBS-TV News. Each day, he and two others—the foreign news editor and the Washington, DC, news editor—make up the list of stories that will be televised that day. It is a list that is submitted only to the executive news editor and anchorman, Dan Rather, for final scrutiny.

If one doubts the excitement in Sturtevant, one need only listen to him as he describes his work, "I want what we all love about this business—seeing things first, watching wars happen, and watching the world unfold. Seeing it firsthand is a treat for me. That's one of the thrills of this business, knowing it all before the people in the street know about it. I like to know things first. Maybe it's a power trip."

Power trip or not, Sturtevant has always craved to be on top of the news—from his free-swinging liberal days as a college student yelling in wild approval as he watched the first draft-card burnings in protest of the Vietnam War to the moderation of his views after a two-and-a-half-month tour of Vietnam conducted by the U.S. State Department.

Sturtevant recalls those days, "In 1966 I was one of thirty graduate students selected by the State Department to tour Vietnam to get a closer look at the U.S. effort there. Apparently Secretary of State Dean Rusk and President Lyndon Johnson felt we college kids really didn't understand what the U.S was trying to do in Vietnam.

"They taught us Vietnamese, gave us access to secret files, and gave us the tour of Vietnam. Most of us had been active in the antiwar movement. I traveled all over Vietnam and in those ten weeks I fell in love with Vietnam as a country, as a story, and as an event. And I knew when I went to work for CBS that I wanted to go back there."

Indeed Peter Sturtevant would fulfill that determination to return to Vietnam. It is an integral part of his total news sense, which had its start when he was a sports reporter on the school newspaper at Nichols Preparatory School in Buffalo, New York. From there his love for reading and writing was enhanced at a Quaker school, Wilmington College in Ohio. During the summers he worked as a cub general assignment reporter on the Buffalo *Evening News*.

"It was at that newspaper that I was really inspired," says Sturtevant. "The city editor, who was very inspirational and my earliest mentor, really could get my adrenalin going when I was in the newsroom. From there it just took off. I became editor of the college newspaper, worked on the yearbook, and was a stringer for the Associated Press in Dayton, Columbus, and Cincinnati. Then I went to the University of Iowa for a master's degree in journalism, where I also served as an AP stringer."

Upon graduation from Iowa, Sturtevant was both certain and uncertain about his future. He explains, "I still was not determined to be a journalist, but I was certain I wanted to be in Washington, where there was journalism, government, and politics. That's where events were happening. I knew in Washington I would find the excitement I wanted. I went there and had interviews with the State Department and on Capitol Hill, but the best offer was at the Washington bureau of CBS as a radio/TV writer."

Having been a magazine journalism major at Iowa, Sturtevant discovered himself in a strange world of television. He explains, "I spent the first four to six months cutting radio tape, listening to committee hearings, overseeing scripts written by White House correspondents. I knew so little. Fortunately, CBS gave me the time and space to learn, and I learned the hard way, by making mistakes and watching what other people did.

"But I think what I did bring to the job was an instinctive sense of what was news, and I enjoyed it. Then I started getting into the TV cutting room and learning how to produce TV pieces and going into the field for news conferences and spending time at the White House and on Capitol Hill—just thinking that Washington was the center of the universe and there was no other kind of news in the world. When I went to Vietnam I felt the same way. Then moving into this kind of domestic news job makes you feel the same way. I guess it depends upon where you are."

It was that feeling of excitement that brought about Sturtevant's assignment to Vietnam. In 1971 he wanted to work in the CBS news bureau in Saigon. He says, "I bugged my bosses for two and a half years, and finally they said okay, though they told me I was still 'awfully young and green.' But I could speak Vietnamese, and most CBS people couldn't. So they told me I had earned my chance. I was there six months as assistant bureau chief; then the chief was reassigned and I was named chief."

That experience in Vietnam deeply affected Sturtevant, who reflects, "Vietnam was a painful experience for the Vietnamese and the Americans alike. It was a very human tragedy. Watching it unfold was a horror, but an experience I will not forget."

His experience in the Saigon bureau eventually led him back to the U.S., and after a short stint as CBS assistant national news editor and northeast bureau chief, Sturtevant was named national news editor in 1973.

In this position as one of the top CBS-TV news executives, Sturtevant has the best of advice for the aspiring journalist. He says, "I was your basic left-wing liberal in college. Now I think I am almost at the point where I have to reach for what I actually think about politicians, public policy, Iran, and El Salvador. I find myself well disciplined, as I rarely think of my own opinion. When I'm asked what I think, I usually have to answer that I don't know and that I'd have to think about it before I could give an answer.

"So I think I have my blinders on to the extent that I'm professional enough not to let my opinions interfere, and I take a lot of pride in that. Often I don't know what my opinions are. Yes, every four years I do vote my convictions, and though I probably have moderated my left-wing politics, I really don't spend much time thinking about that.

"This kind of work is of such overriding importance and interest to me that I don't spend much time thinking of my own opinions. I suppose I try to exercise my news judgment on the basis of the sum total of my experience."

Peter Sturtevant is a journalist—a journalist who discovers an excitement in his every task. He demonstrates this in a final thought, "I really enjoy journalists as a group. They are interesting people, more intense and curious than others. I love the excitement in journalism. People and events are stimulating—both intellectually and emotionally."

Chapter IX

RICK BROWN—"I'M ALWAYS COMPETING AGAINST TWO OTHER GUYS."

Photo by Rod Vahl

Behind the scenes and calling the shots—that's the position preferred by many broadcast journalists, and Rick Brown is no exception. The young Missourian in his 30's sits in a cluttered office in the CBS building on the Near North Side of Chicago and commands a small company of reporters, producers, editors, and camera crews who daily spread out over a twelve-state Midwestern area to cover both the breaking news of the day and light news features for the daily CBS-TV network newscasts originating out of the New York City studios.

Rick Brown's title is chief of the Chicago News Bureau for CBS-TV News, one of eight national news bureaus covering the United States under the direct command of Peter Sturtevant, national news editor for CBS-TV News in New York. And Brown loves his work: "I think the greatest satisfaction I get out of this job is when I think I'm doing well. That is, when it seems that everything I've ever learned is making my job work for me and getting me a good story."

It was in high school that Brown decided he wanted a journalism career, though it was not necessarily a television news career that he thought of at that time. He explains, "I was sports editor of our high school newspaper at University City High School in University City, which is a suburb of St. Louis. I loved sports and enrolled in the school of journalism at the University of Missouri with the idea of becoming a sportswriter."

But that changed, as he explains, "Between my junior and senior years at Missouri I had a summer job at KST in St. Louis as a newswriter, and it was then that I decided I wanted to get on the news side of broadcasting."

After graduating in 1970, Rick Brown landed a job in Quincy, Illinois, on WGBN, an NBC affiliate radio-television station. "I did everything there," he says, "I covered stories in the field, I did anchoring. You name it, sports, weather, editing, writing. Even some disk jockey work."

But Brown was itchy for a different location because he was not happy with the operations at the Quincy station. During the year he was there, he took his free days in the middle of the week and drove to other cities to check out opportunities. This led him to Milwaukee at WITI-TV, where he accepted an offer as a general assignment reporter.

He moved fast at the Milwaukee station. He explains, "After two years as a street reporter I decided to go behind the scenes rather than be on the air. I took over the weekend news desk as a producer, assignment editor, and writer. That lasted a year, and then I produced the daily 6 and 10 p.m. newscasts. Then I was made news director."

In 1977 the station switched from ABC to CBS, and this move impressed Rick Brown much. He explains, "In 1977 CBS was much better than ABC in news. We would hear from the news bureau chief in Chicago every day, whereas we rarely heard from ABC. Well, I developed a good rapport with the bureau, and in 1978 I was offered a TV news producer's spot with the CBS Chicago news bureau. I took it, and suddenly I was out on the streets again, producing stories for the morning news."

Being in the right spot at the right time pays off, and it paid off for Rick Brown, for when the bureau chief was promoted, Brown was named chief.

Brown's staff includes five reporters, three producers, three editors, and three camera crews of a cameraman and a soundman. In addition, Brown maintains a list of free-lance reporting crews he can call upon at any time. "We cover twelve states, and sometimes we need to call these free-lance crews because our own people are tied up or because the free-lancers are closer and more knowledgeable about the story we want," Brown explains.

Sometimes the CBS affiliate stations throughout the twelve-state area are used. Brown explains, "There are some who can really do the job, others that can't. We usually know who can be relied upon to come through for us."

Brown's workday starts in the morning with a telephone conference that includes the national news editor in New York City and the eight bureau chiefs. Stories are initiated either in New York or in the bureaus; all are followed up by several telephone conferences a day between the New York news office and the individual bureaus.

Brown explains, "Every day we get a preliminary lineup of the news for the CBS Evening News. We send our reporters and producers and camera crews out. An editor writes the script and transmits it to New York. They edit it and make a decision. Once we have approval on a story, we edit the entire story and film and feed it to New York between

4:30 and 5:30 p.m. If the Evening News doesn't use the story, it usually goes on the Morning News."

All in a day's time! One wonders how such stories can be done sometimes within a day, and Brown explains, "Logistics is a big part of my job. We'll fly an entire crew to Omaha and back within hours for a story. There are airline flights to schedule, rental cars to hire, sometimes motel reservations to make. We even charter planes to take the crews to locations that are not served by the commercial airlines."

But not all the stories are breaking, on-the-spot news. It is the bureau chief's responsibility to develop enterprising light features. For instance, Brown read on the wire services about an Ohio farm family with four children under ten years of age who were all geniuses. He sent a crew to Ohio to develop a feature for the national newscast.

Brown says. "We constantly read the wire services, read out-of-town newspapers, read newsmagazines, call affiliate stations to see what they are doing. We depend upon all these sources for potential stories for the network newscasts."

That is the toughest part of Brown's job—making value judgments on news. He says, "We constantly decide what is newsworthy or not. It's tough to decide what to do. Can we get a good story? Is it of interest to a national audience? What is the expense? If it's a hard news story, does it need to be told? If it's a light feature, is it cute enough?"

And there is the competition that Brown must be conscious of every hour. He explains, "I'm always competing against two other guys—the ABC and NBC bureaus. That's tough, because every day I am compared to what two other guys are doing. Are they doing the same story? If not, do I have a good story? If so, do they have it before I do? There is always the competitive aspect to this television newscasting."

Brown believes television newscasting is far more difficult than newspaper reporting. "Time is so limited for each story. We have only a few minutes, sometimes less. And then we have to write the story to the film we have. Things can go wrong. If the film gets screwed up, the story is killed. We just have to cram everything into a short period of time. That's frustrating, but it's part of the job."

One of Rick Brown's responsibilities is to hire reporters. He has his standards for the reporter, and they are demanding. "I want the reporter who is well-read, one who reads newspapers, magazines, and books. The reporter needs to be well-rounded. I don't like the reporter who has to be filled in on everything. He needs to know who people are and what's going on at City Hall. I want a well-educated reporter. He needs to write well."

And then there are the attitudes he seeks in the reporter. "The television reporter needs to recognize the need for speed because of the deadlines, so I want an aggressive and competitive reporter."

Although Brown majored in journalism in college, he does not believe that the journalism major is essential. He says, "I would recommend a very broad liberal arts bachelor's degree. The reporter needs great knowledge in history, government, economics, and literature."

Brown looks at both the high school and college student and explains, "If I were molding a student who is bright, I would have him read all the history he could read. How can a reporter cover news events if he doesn't understand them? If you know history, you know the significance of what happens. There is so much happening that has historical precedent. If you know the history, you have a better perspective of what is happening now. When you study history you realize that things that happen today happened many times before in similar form."

Brown emphasizes the need for the reporter to comprehend the functions of political science. "Know how government works! Make yourself broad in knowledge, because that's what you'll be covering as a reporter—everything. How can you cover anything if you don't know anything?"

The news bureau chief believes that much of the technical side of television news broadcasting can be learned on the job. "That comes fast," he says, "but all these other things—business, economics, government, politics, literature—they don't come as easily."

Thus, Brown recommends a double major in journalism and in economics, political science, or business. Or he believes one might obtain a master's degree in journalism and devote a bachelor's degree to other areas.

"It's so important to be well-rounded. As I look back, I don't think I learned enough in high school or college. I do more reading now. I just wish I had studied more literature, history, political science, and business and economics. That's what the reporter covers."

Once again Brown emphasizes the attitudes. He says, "The toughest part is knowing when to go on a story. You're in competition every day, and you've got to be aggressive and competitive. There is a limited number of people in this business, but it's highly competitive. Often there are long hours and six- and seven-day weeks."

Obviously, Rick Brown craves television broadcasting. Not only is it evident in his approach to his job as a news bureau chief, but it is well demonstrated in what he says—"The nitty-gritty of the job is covering stories. I like best of all doing a story, doing it well, and seeing it go on the air."

Chapter X

FORREST RESPESS—"I'VE GOT TO HAVE A LITTLE PIZZAZZ!"

Photo by Rod Vahl

Good morning. Marty McNeeley for WGN News. A political surprise Wednesday night when it was announced that 13-year-old Walter Polovchak's parents have returned to Russia—without Walter. Canadian air controllers return to their jobs, but other foreign controllers may start boycotting American flights. And on Chicago's baseball scene, a double dunking. But good news, the Sting posts its 21st victory for the year. These and other stories on "Nightbeat."

Those were the lead-in headlines broadcast by the newscaster for a 1 a.m. news program on station WGN-TV in Chicago, and the person responsible for the total production of such a news program is called the producer. It is his responsibility to assure that the thirty-minute program is all in order, with the commercials, weather, sports, and news all prepared for the on-camera talent to present to the viewers.

Forrest Respess is a long-time news producer for WGN, a job he loves with a passion. "I like the unexpectedness. You never know what's going to happen. Nothing is predictable in this business. A lot of people out there look at us television people as a bit ghoulish, as much of our existence depends upon problems of people—their tragedies, shootings, plane crashes. You can never expect routine. I like that lack of routine. And you meet interesting people. We have a parade of celebrities coming through these studios."

Like most veteran broadcast journalists, Respess has a wide background of experience that followed his graduation from the University of Cincinnati with a bachelor of arts degree in radio education. He values his first experiences as he roughly traces his initiation into radio, "I was in Lancaster, Ohio, and I was a disc jockey, newsman, man on the street, announcer, and program director. I was there three years, and I did everything. It's so important to work on a small station, because that's where you really must work at every job that exists."

From Ohio, Respess moved to Indianapolis, where he worked as announcer and continuity director. When television started, he went to New York City to study stage management and television production at the American Theatre Wing.

For the next few years he worked on television stations in Indiana, Illinois, Missouri, and Michigan, finally ending up at WGN-TV, where he was soon promoted to producer for the "Nightbeat" news show.

The varied experiences in those jobs well prepared Forrest Respess for the extremely competitive television newscasts of Chicago, but he finds it a challenge. "I think what it is during the course of the evening is that you're putting something together. It's like an assembly line. You have all these various pieces and you're putting some kind of a jigsaw puzzle together. A lot of it is mechanical. You have to put your tapes, slides, and films together. As you get the show on the air, you are into a production. It's show business. You've got a show to put on the air. You're dependent upon everyone. Your director, your reporters, your engineers. Once it's on the air, it's out of your hands, but it's really your baby when you have to get it together. You call the shots. And then you turn it over to the talent and just hope everything comes out as you planned it. And when you're finished and everything comes out right, you really feel good about it."

As a producer, Respess starts his daily job at 6:30 p.m., and the first task he faces is to study the log, a time schedule for his newscast that will include:

Commercials: 6½ minutes
Sports: 4½ minutes
Weather: 45 seconds
Opening and closing: 1½ minutes

Respess explains, "Those are my fixed figures. That leaves me 14 minutes and 45 seconds for news. But sometimes I find that sports runs only four minutes, so that gives me an extra 30 seconds for news. On this particular news show, I don't have to sit and time it with a stopwatch. We're on all night, and I can run over a minute or even be short. I can make various adjustments like pulling out a promotional bit or a public service announcement."

After noting his time allotment, Respess reviews the news shown during the day and checks what the nine o'clock newscast is presenting. He says, "I also look at the national wire news to see what stories are breaking. In this job you're really working twenty-four hours. You're always doing something that prepares you for the job. You're listening to the radio, watching television, reading the newspapers. So when you come in to work, you almost know what your lead story will be."

The news producer is in constant contact with what the reporters are doing, but he recognizes that the unexpected can occur at any time. He says, "Everything can break loose, and that's something you have to be prepared for so you don't panic. One night we were on the air and the overnight radio girl called the control room and said the new pope, John Paul I, had just died. I came downstairs from the control room and went in and hacked out a brief news item and took it to the newscaster and told him to read it and to say we'd have more details later."

Immediately Respess needed to gain more information, and he explains, "I came down and fortunately knew where the tape of his installation was and also a tape of one of his first talks. We had a videotape of it. I got that and took it to videotape, and before the show was off the air, we had a pretty good package for it."

Just what is a good news package for the viewers? Forrest Respess is constantly seeking a good balance of local, national, and international news as he plans the "Nightbeat" newscast. He says, "I can recall a very recent newscast when we were overbalanced with local stories. "Nightbeat" is supposed to be an all-encompassing show, yet there just were no strong national or international stories. There were some sidebar and feature pieces I could have used, but they wouldn't have fit in with the content of the rest of the show. If I'd thrown in a Berlin Wall feature, I would not have had anything to relate it to."

Respess constantly is aware of the audience. "A well-balanced show is one that doesn't confuse the audience. It's one that doesn't give an overdose of journalism. I've always contended that we're still show business and we've got to have something appealing that people will want to watch. You can have the most perfect journalism show in the world, but if it's dull and dry, there's no point in doing it because you won't have an audience."

In the writing of that newscast, the WGN producer says, "I've got to have a little pizzazz! When I'm writing, I try to use different writing techniques, such as if I can find some alliteration that the talent can handle. But if the talent that's broadcasting is more journalistically inclined, I can't do it because it would fall flat. I usually need to write pretty much a straight newscast. I'd like to zing it up more."

For "Nightbeat," Respess puts that little zing into the news as much as he can. He rewrites the script from the earlier broadcasts, and as producer he exerts full control over the writing. He says, "I don't cater to a talent's likes and dislikes. Sometimes I write something and then hear something over the air that I didn't write. I don't mind changing a sentence to make it easier to read, but when a talent changes the meaning without telling me, I do get upset."

But that "zing" Respess sometimes uses must never be done at the sacrifice of being objective, and he recognizes this as he says, "You have to lose your subjectiveness of the news and be as objective as possible. Whatever the person, whatever the cause, you must consider every story on its news value. It's hard to look at a story and do this. But you must look at the effects the story will have upon the viewers."

As he looks at the list of potential stories for the broadcast, Respess says he attempts to determine which are the best stories and which the station should be reporting. He explains, "You can't overweight one story. The newscast must flow and be interesting. Surveys show that an audience hears about one-third of what is said. They won't even hear that one-third if the show isn't interesting."

College training is a must in the view of Respess for those who aspire to a broadcast journalism career, and he recommends a strong background in English, political science, structure of government, and especially knowledge of civil and criminal law. He notes, "A prospective journalist should know enough about law so that he understands the legal ramifications of a trial. It's important that the reporter can intelligently discuss events with a solid understanding of history."

Respess also recommends a strong background in writing, as he explains, "So often the producer must be a writer because, as in my spot, I do a lot of writing. I have to catch my own errors. And that writing ability must extend into many areas—news, sports, weather."

Then, looking at station personnel, the veteran producer states, "If you're lucky, your first job will be a low-paying, long-hours, hard-working job in a small market where you go out on the street with a camera and microphone. You'll go back to the newsroom and write your story, you'll edit your own story, and then you'll go into the studio and read your own story on the air. If you do all those things, you'll get to know the functions of all the people with whom you work. You'll appreciate their duties—the announcer, engineer, cameramen, switcher, director, artist, projectionist. You'll understand what they are contributing to your show."

Forrest Respess has a final word for the aspiring broadcast journalist as he quips, "It's no pipe and slippers routine, but it's excitement on a daily basis!"

Chapter XI

JOAN LUNDEN—"I LEARNED AND MADE MY OWN MISTAKES RIGHT ON THE AIR."

Photo Courtesy of ABC-TV

"Being a woman, you're expected to fall flat on your face. They wait for you to fall flat on your face, and they can't believe it when you don't." That is how Joan Lunden, reporter-interviewer on ABC-TV's "Good Morning America" with David Hartman, describes some of the hard-nosed newsmen who looked upon her as simply another "pretty blonde" propped before the camera so dear old dad can have a bit of sex appeal with his bacon and eggs every morning.

And Joan Luden understands those skeptical TV studio journalists. After all, she had not studied journalism in college, nor had she harbored a single dream of being a television personality. In fact, Joan Lunden did not have the slightest idea what she would do upon graduation from college nine years ago.

"I just studied and traveled after I graduated from high school in Sacramento, California, when I was sixteen years old," she explains. Upon enrolling at the University of California, Lunden immediately toured most of the globe through the World Campus Afloat program, and upon her return she decided she would rather study in another country. That decision led her to study Spanish and anthropology for three years at Mexico City's Universidad de Las Americas.

After returning from Mexico, Lunden enrolled in college to complete studies for a liberal arts degree. "And there I was," she says, "with a degree, a lot of traveling and different schools behind me, and not knowing what I would do to make a living."

However, one of Lunden's friends, who was an advertising salesman for Sacramento's KCRA-TV, suggested that she talk with the station manager, who wanted to add a woman to the news broadcasting staff.

"My whole realm of thinking didn't allow me to think about a television news career," explains Lunden, but since she had no definite plans for the future, she visited the station. She says, "They were interested in me because of my liberal arts and traveling background, and they said I could start as a consumer reporter."

But how does one become a consumer reporter with no background? Well, if one wants a job badly enough, one educates oneself, and that is precisely what Lunden did. She explains, "I went to bookstores and bought all the books I could find on consumerism. Then I went to California State University and found a professor in consumerism and asked him if I could spend a few hours a week with him. He agreed, as he was in favor of TV consumer coverage. Giving me a reference list, he sent me to the Department of Agriculture and the Department of Consumer Affairs at the state capitol to acquaint myself with the issues. Then I talked with the legislators and consumer advocates to learn what they were doing. It was a matter of finding the people who knew about consumer news and what the issues were."

It was strictly a task of education, Lunden says. "I'm totally self-educated in TV. I had to learn editing and writing on my own. I learned and made my own mistakes right on the air. And I had a lot of criticism for those mistakes. Today you can make your mistakes in college where you learn to write and to produce TV shows. But I had to make them at the station."

Six months later this self-education brought the young Lunden a promotion to producing and anchoring the noon news at KCRA-TV. She would arise at 5:30 a.m. to watch the news on the "Today" show, selecting network inserts for the noon broadcast. Then she would scrutinize the wire copy, deciding what stories she wanted to use. After preparing the copy and slides, Lunden would consult with the news show director to discuss the program.

"It was hectic," she admits. "After consulting with the co-anchor on what he was to do, I would take my coffee, sit on the news set, and do the show. During commercials, I would phone the director to check on how our time was to see if I had to cut or use an extra story. Sometimes you produce right from that anchor chair. It was wonderful because you learned. I did everything, so when I came to New York I had an empathy for everyone. With that kind of experience you have a sensitivity to all those things, which, I think, is better when you start reporting."

Nearly two years passed for Joan Lunden and her KCRA-TV job, and she had yet to go out onto the streets to do the reporting chores of a journalist.

But in 1975 that changed.

A television consultant firm was called to Sacramento to help a rival TV station increase its ratings. In that process, the consultants spot top personalities of other stations and tape them, to help them find better jobs elsewhere in the nation. Joan Lunden was one of those identified and taped, and soon she was being offered jobs in cities such as Detroit, St. Louis, and Milwaukee.

She says, "But I didn't know much about the TV industry, and I had no way to assess these offers. Well, the same friend who suggested I talk to KCRA-TV now suggested that I call a friend of his at ABC-TV in New York for advice on which job offer to accept."

Little did she suspect that ABC was looking for some new people for their five ABC-owned TV stations in New York. ABC asked her to send them a tape also, and the immediate result was an offer to be a reporter and weekend anchorperson on WABC-TV in New York City.

Of course, the skeptics were vocal once again as they asked, "How can you go to New York City and be a reporter when you've never been a street reporter?"

But Lunden had her reply, "The worst that can happen is to fail and come back to Sacramento. One has to take certain chances in life. They said they'd train me. Well, I was trained in Sacramento, and there was no reason to believe I wouldn't be trained in New York."

However, little training was offered to Lunden once she was in New York, and on the third day of work she was assigned to do a story. She remembers that first reporting assignment well. "Here was a little sheet off the wire about a bombing and conspiracy trial of one of the flower children who had bombed recruiting stations. I had never been in a courtroom, let alone the Supreme Court. I had to go alone. No file crew—only my mike and my recorder.

"Inside the courtroom these kids were jammed all over the room. I looked around and saw the front row with all these people with their little notepads and said 'That's my place!' I found a seat and watched what everyone did. My only instructions were that if the defendant was found innocent, I was to stick with her like glue and get outside with her and the camera crew would find us. Well, she was found innocent. The place went up for grabs because all those kids were there. I followed her outside and got a sound mike with her. It made network news."

However, that wasn't the end of Lunden's initiation into street reporting. Arriving back at the station, she was handed another assignment. Lunden was sent to the U.S. Mission at the United Nations headquarters where Secretary of State Henry Kissinger was to speak to the leaders of the OPEC nations concerning the oil crisis.

Lunden recalls once again, "There was a big demonstration against Kissinger right across the street, because they didn't feel that he should be dealing with OPEC nations. My instructions were to stay outside, see if I could get to talk with Kissinger, and get the demonstrations going on across the street. Hundreds of network press people came by me, all going inside and having their cameras checked by the Secret Service. I was the only person standing out there by myself with my tape crew, and I had never worked with video tape before. And I'm thinking something must be wrong, standing there by myself.''

Then Kissinger arrived.

Lunden reminisces, ''I knew when he arrived. There were Secret Service men all over. He got out of the car. Now, I'm the only press person left. The Secret Service is like five deep around him. He's coming up the steps, and my heart is pounding harder and harder. I finally just said, 'Mr. Secretary . . .' And you know, when the television lights go on, it really does make a little path. It's like opening up the waters and you walk across. The Secret Service moved to stop me, and Mr. Kissinger just turned around and said, 'I'll talk with the young lady.'

''I thought to myself, 'Joan, you better have a question to ask!' So I asked him what were his feelings about the demonstrators across the street. He didn't really say anything. It was a very nebulous, meaningless answer, but the fact was that he made a statement. And it was used as network news that night and in the morning. That was my first day on the job in New York.''

Having learned reporting at a rapid pace, Lunden was soon regarded as one of the brightest personalities on New York TV. After a year, she was spotted by executives of the ''Good Morning America'' show on ABC, who were looking for a woman to report on new products, new trends, and new life-styles.

Lunden explains, ''They asked me to do a monthly report. Well, I had prepared for consumer reporting back in Sacramento, so why not new products? ABC liked the spot on the show, and the job soon went from once a month to every other week.''

ABC then conducted studies on Lunden's appearances and found that viewers liked her, and soon she was doing film pieces for the show. The fast-rising reporter says, ''It grew to where I was working 18 hours a day. I would go to ''Good Morning America'' at 5 a.m., rehearse, do the show, and then run up the block to WABC-TV at 9 a.m. to get my assignments, be out, and then back to do the news show at 6 p.m. that night. And sometimes I had to come back and do the 11 p.m. show.''

This schedule increased when Lunden became the substitute on ''Good Morning America'' for Sandy Hill, but as Lunden says, ''It was getting a little hairy!'' She could not do only the ''Good Morning America'' show, as it was not a full-time job. Nor could she afford to quit her full-time job at WABC-TV. And yet she did not want to surrender the network show.

This hectic pace was coupled with two realities as the decade of the 80's emerged. She and her husband, TV producer Michael Krauss, were expecting their first child, and Lunden's contract with WABC-TV was due to expire. Again, being a woman unafraid to take chances, Lunden approached the "Good Morning America" executives and sought a full-time job as a reporter and substitute co-host for the show.

It was perfect timing for the 30-year-old journalist, as Sandy Hill wanted to be relieved of the in-studio chores, and the executives thought it was "great" that Lunden was pregnant. She explains, "The bosses loved my pregnancy. They felt that instead of being just a cute blonde on the show, I would be the warm, glowing mother. Yes, being pregnant helped my career! The response of the viewers was incredible, as they could relate to me."

The rest is history for Joan Lunden, who is just entering her second year as reporter-interviewer on "Good Morning America." She is "cute" indeed. But she is much more—a self-educated journalist who is willing to take chances despite the criticism that often prevails at every step of success.

For aspiring newspeople, she has good advice, "Criticism will happen at every level. When I went to KCRA-TV other reporters looked upon me with disdain, wondering how the station could have hired a person who had not worked at twelve newspapers and radio stations before being hired. The same thing happened at WABC-TV in New York. The only way you get over it is to stay with the station, work hard, do your job, and earn their respect. It's the only way to fight criticism. At first, it's not easy. But if you are on time, if you tell the story and do it right, and if they see you do this consistently, they gain respect for you."

And Joan Lunden has earned that respect.

Chapter XII

JOHN DRURY—"YOU HAVE TO HAVE A REGARD FOR PEOPLE"

Photo Courtesy of WGN-TV

High school—certainly that is the time when students dream. Dream of tossing 40-yard touchdown passes for the New York Jets. Dream of bowing in front of glaring footlights in a Broadway musical. And, yes, dream of sitting before television cameras and telling the viewers of fastbreaking world events.

It was no different for John Drury, the popular news anchorman for Chicago television station WGN-TV. Drury reminisces, "When I was in high school I knew I wanted to be a broadcaster. I would listen to Edward R. Murrow's broadcasts from Europe in the 30's with the rise of Hitler's Germany and his reporting from there. I was always fascinated by it. Somewhere in the back of my mind, I knew I wanted to be an electronic journalist."

While he attended high school in West Aurora, Illinois, John Drury wrote anything he possibly could. Writing was important to him, he explains. "I wrote much in high school for the school newspaper and I always wrote. I don't know if I was always interested in being precisely a reporter, but I've always been interested in writing. I've always been a storyteller, which is what journalism essentially is, I think."

With both a love for writing and a passion for the Murrow broadcasts, it was only natural that Drury should finally say to himself, "I want not only to write it, I want to say it."

That double desire led Drury to Lyons Township Junior College in LaGrange, Illinois, for two years and then a transfer to the University of Iowa, where he majored in speech, which he thought was the best route to a broadcasting career. While there, he worked on WSUI, the university's educational radio station, broadcasting both news and sports events.

While at the university, Drury also worked for the commercial station in Iowa City, KXIC, where he performed most radio duties, including reporting, newscasting, and sportscasting. "Though I majored in speech,

I also enrolled in some journalism courses so I could do the news functions of radio as well as announcing and sportscasting,'' he explains.

Before he could graduate, Drury found himself broke, so he immediately entered broadcasting with a job at KSTT, a small radio station in Davenport, Iowa. ''I was there only a couple of months,'' he explains, ''and I moved to Fort Wayne, Indiana, where I was a staff announcer. But I soon moved to another station where I could go out and report and write the news and then put it on the air. The whole thing!''

Drury finally was doing what he always wanted to do—telling the world just what was going on. He left Fort Wayne and moved to Indianapolis and then to WTMJ in Milwaukee, Wisconsin, where he stayed for seven years.

Finally he reached Chicago, where he has been for more than twenty years, working first for WBBM-TV, then WGN-TV. Chicago is a competitive television market, and Drury was one of the most popular newsmen and anchormen. Thus, he was lured to WLS-TV for ten years. A few years ago he returned to WGN-TV, where he is now the top anchorman.

Drury emphasizes the need for aspiring reporters to recognize that it is vital to gain experience on the smaller radio and television stations, because ''that's where you gain all the different kinds of experience needed.''

He says, ''Walter Cronkite spent a number of years doing many different things. In fact, he was in the newspaper business at first, but he wanted to do it another way. I wanted the same thing—to communicate verbally. This television business is both writing and speaking, but it isn't a writer's medium. There's more editing to it, more selectivity. Writing is more transitional. The daily writing of a television reporter would be about one-fifth of what a newspaper reporter writes every day.''

Drury is in a position to judge the new reporters coming out of the colleges, and he likes them. He says, ''Today's reporters are better prepared. They know the industry and they know what's going on in the world. They know how to write, they know how to talk. They have a deep-down desire to communicate, and that's what this is all about.

''I like especially those new reporters who want to create things. What I like is what I see in many of them, a social conscience. A lot of these kids went to journalism school because of Watergate. They really feel they can reform things and change the world, and I think that is one of the goals of journalism—to reform.''

However, John Drury also sees the other kind of journalist, the one seeking fame and fortune. He says, ''Many of them are attracted by the money. When I started I just wanted to broadcast. I didn't worry about the money. Those I see who are motivated by the money aren't as good as those with a social conscience.''

Thus, Drury recommends that students become interested in history, economics, government, world affairs. It is through those studies, he believes, that young, aspiring television journalists develop a keen sense of world events and how those events affect people all over the world.

The desire to communicate is vitally important, Drury feels, and he is constantly evaluating himself. "I know if I've done something good or bad. The thing I like most to do is to go out on the street and report, not what I do here in the studio as an anchorman. In television you have to find the people who can tell the story. And that has to be done within a few minutes. We don't have the luxury of the newspaper with all its space. Our time is short. So it's important to be able to go out on the street, talk with the people, and come back with a story that can be told in a short time on the screen."

To do that, John Drury once again emphasizes the need for social conscience, and he explains, "You have to have a regard for people. You need to possess a sense of what is right and what is wrong. You need a sense of integrity so that you are honest in what you are doing and telling the truth as best you can."

He adds, "You must acknowledge that you have a whole set of prejudices, and you must know what they are. Yes, sometimes you have to take a side. Sometimes you have to say something is wrong and you must do something about it. In investigative reporting I see things in this city that are wrong, and you have to raise your moral indignation over these things and go after them. You must try to ferret out those who are cheating the government, cheating the taxpayers. There's a lot of wrongdoing in government that should be sought out. You need to develop sources who will tell you what's going on and what's wrong."

It isn't easy for a journalist such as John Drury to do investigative reporting, for he is a man with much compassion for all his fellow beings. This is evident as he says, "I find myself feeling very sympathetic with even the person who is doing something wrong. I hate to see anyone go to jail."

But Drury knows his responsibilities and says, "You have to pursue it. Advocacy journalism is not in vogue now. It was true in the 1960's and 70's when there were stands against the Vietnam War and against the abuse of blacks. But we've shifted today to a different set of values. We got out of the war. Reforms were instituted in the South."

Drury discovered that he could be selective about issues. He says, "I was against the Vietnam War. It wasn't productive, and it was just a loss of a lot of lives. I was for the civil rights movement in the 60's. I think the journalist generally should be avant garde—look for trends that are changing society."

He is serious as he says, "One of the great indictments of journalists is that they are too middle-class. They preserve the middle class rather than anything else. They tend not to pursue new ideas or they step on new ideas that are really the ideas that are taking form and changing society. I think reporters should be many things. They must be as fair as possible."

Probably what John Drury dislikes most about electronic journalism is the competitiveness among television stations. He says, "Ratings are everything. We have no control over this. Oh, I don't really mind the competitiveness, but it's too bad some good people fall by the wayside. It's the cosmetic thing. There is an electorate out in that audience. It's just like politics. If you don't please the electorate, you're finished in television."

He continues, "The trouble is that a person can manipulate, as Jimmy Carter did with the people on television. But the company knows what the public wants. That's what is called market journalism, and that's bad."

Drury illustrates the problem by noting a recent change in CBS-TV news. He explains, "I don't know how you separate the cosmetic thing from good journalism. CBS made a decision upon the cosmetic thing. Both Roger Mudd and Dan Rather are excellent journalists, but they went with Rather, solely upon the basis of charisma. They went with Rather because they thought they'd build a bigger audience with him. It was strictly a cosmetic decision."

And the "show biz" element pops up again, with Drury saying, "There's a great deal of theatrics in this business. Walter Cronkite was a performer. He always loved covering those space shots—anything that was important to the American people. In anything, he was always a quick study and well prepared. Cronkite may not have been a great thinker, but he was the best communicator, and that is really what it's all about."

Indeed, reporters will fail in achieving a successful television broadcasting career, and Drury says, "Those who fail are not honest. There are those who will do anything for their careers. I've seen reporters do stories in which they took great liberties to assume things that didn't happen just to enhance them and make them more interesting."

Looking to the future, John Drury sees many opportunities for young people in broadcast journalism. He says, "There are a lot of ambitious people in this business. They are high-energy and high-drive people. Get yourself a broad background, an interest in everything. If you want to specialize, there's room for areas such as science reporting. But be certain to get that broad background, mostly in the areas of politics and government. Understand the English language. I don't think a journalism major is necessary. Just get the background you'll need to get out on any story assigned to you."

Perhaps John Drury can offer the best advice in simply saying what his work means to him: "I like best the variety in my job. The different things that happen everyday. The challenge. I like best to go out on the street and do a story. It's the only thing that keeps me here. It's being out there with the people who are making the news."

Chapter XIII

MILO HAMILTON—"I FELT THE WHOLE WORLD WAS WAITING TO HEAR FROM ME."

Photo Courtesy of WGN-TV

Doctor. Actor. Policeman. Engineer. Lawyer. Go into any high school across the nation and you will discover at least one student who knows exactly the career he or she wants.

And that's the way it was for Milo Hamilton, the Chicago television sportscaster for the Chicago Cubs baseball team and the DePaul University and professional Chicago Bulls basketball teams. Hamilton says, "I lived in a small town, Fairfield, Iowa, where you've got to do everything. I played all the sports, and I did a lot of acting in school plays and a lot of singing. I just figured I'd do something with my voice because I was blessed with a good voice."

The sports Milo Hamilton played were football, basketball, and baseball, but as he explains, "I knew I wasn't good enough to play in college or the pro field because I was always a borderline athlete. Oh, I could have played basketball at a small college, but I just didn't want to."

But that love for sports and his good voice led Hamilton to one conclusion—"I knew that sportscasting would be for me. I felt the whole world was waiting to hear from me."

Upon his graduation from the University of Iowa with a major in speech and radio, Hamilton started at the grass-roots level with a sports broadcasting job for radio station KSTT in Davenport, Iowa. The city is a part of the Quad Cities, a metropolitan area of Iowa and Illinois cities separated by the Mississippi River, with a wealth of sports activities at more than ten high schools and three colleges as well as minor league professional baseball.

Practically every professional journalist advises beginners to gain as much experience as possible on a small newspaper or radio or television station, for that is where the journalist can gain a breadth of experience. It was true for Milo Hamilton, as he explains, "The first month on the job at KSTT I did more than thirty basketball games. I did high school and college games and also the Quad City Blackhawks, who are now the Atlanta Hawks in the National Basketball Association. You might say that I went to a perfect place to start and get my master's degree in broadcasting."

That experience for Hamilton included over 100 basketball games and 40 football games a year. He also covered the University of Iowa's Big Ten games and the minor league baseball team in the Quad Cities. In addition, Hamilton broadcast over 100 boxing matches a year.

Laughing, Hamilton says, "The joke in Davenport in those years was that if someone threw a ball in the air, I'd be there to cover it."

After three years in Davenport and the minor leagues, Hamilton successfully auditioned for his first major league broadcasting job, in St. Louis, where he covered the old St. Louis Browns baseball team. When the Browns moved to Baltimore to become the Orioles, Hamilton switched to working with Harry Carey on broadcasting the St. Louis Cardinals. He continued his interest in basketball by covering the St. Louis University games as well.

It was not all roses in St. Louis for Hamilton. Only a year after he joined Carey, the top brass decided they wanted a former major league baseball player to help Carey, and Hamilton had to step aside for Joe Garagiola.

Hamilton wouldn't permit such a disappointment to thwart his ambitions. "I heard about an opening at WIND in Chicago to help the famous Bert Wilson do the Cub's games. I called, went in the next day, and got the job." He was there eleven years, doing the Cub's games first and then broadcasting the White Sox team after the Cubs switched to another station.

The next step up for Hamilton was a ten-year job with the Atlanta Braves, where he experienced one of the highlights of his career, the broadcasting of Hank Aaron's all-time homerun record smash. From Atlanta, Hamilton moved to Pittsburgh for the Pirates, and finally back to Chicago to broadcast the Cubs' games again with Jack Brickhouse and now Harry Carey.

Although Hamilton enjoys baseball more than other sports, he loves all sports, and he says, "I think to become a major league announcer you must really be interested in all sports. You soon find out that although you still enjoy sportscasting, there are aspects that take away some of the glamour. The traveling is torture. You work a lot of hours that people don't realize you work. It's like a big league ball player; it's tough to get to the big league as a player, but it's tougher to stay there. The same is true for the announcer, because there are a lot who want your job."

To stay on top, the sportscaster must constantly do his homework, and many sportscasters have lost their jobs or failed to climb higher simply because they overlooked this "must" of the game. Hamilton learned this early, and he explains, "I keep a lot more records than some broadcasters do because I don't like generalities. If Pete Rose is hitting .500 against the Cubs, I would rather be able to say he's hitting 12 for 24 than to say he's having a good year against the Cubs."

Hamilton emphasizes, "You've got to be different every day, not repeating the same things game after game. You can't keep trotting out the same act. Your style can be the same; that's your identification, that's how people decide if they like you or not over the long haul. But the content of what you're saying must be good and diversified. You can't keep referring to the good old days, because a lot of people watching you don't remember those days. You have a whole new audience every day. Sometimes you can refer to Lou Gehrig or Babe Ruth because your senior citizens get a kick out of that, but that can't be your whole act."

There is a special discipline in television sportscasting, and it is very different from radio broadcasting. Hamilton explains, "In the old days on radio you could ramble on, but on television the people see it. My job is to recap the action two or three times during the game because people come and go. They catch the game and you in the third inning or the seventh inning, or they miss an inning or two. So you constantly let the fans know what's happened during the game. A lot of them keep a scorecard and check it against what you're telling them."

That discipline also requires the sportscaster to make the game a good game. That may sound strange, but Hamilton elaborates, "I want every game to be better than yesterday's. Some games don't lend themselves to broadcasting. You do 150 games a year and some of them are bound to be bummers no matter how hard you try."

That's when some broadcasters seem to like to tear into the players. Indeed, the sportscaster is a critic as well as a reporter, and though Hamilton does not shun such responsibilities, he is concerned about his credibility. He says, "There's a way to maintain your credibility as a critic and still not ruin players. My philosophy is that if it affects the game and the team is losing, you have to report it; if the fans are sitting there and you ignore the bad things, they'll wonder if you're seeing the same game."

He continues, "I'm critical, yes, but I draw a line and don't belabor it. After the game is over and I've recapped it, I forget it. Let's say a player makes an error in the fifth inning; it should have been an easy play, but he throws the ball away and the other team scores three runs to win the game. I have to mention it at the end of the game because it definitely affected the outcome. But that's the last you'll hear of it from me. The next day it's not mentioned. People get tired of hearing about it, and it isn't fair to the player, who feels worse about it than you do."

Often both local and national network news shows are pegged as show business, and there is no doubt that pure journalists are quite critical of the theatrics that has become a part of broadcast journalism. The same is true of sports broadcasting, as Hamilton acknowledges, "Theatrics is there because you're in show business. If you listened to tapes from the 1930's and 40's, the baseball broadcast would sound a lot different from what you hear today."

Looking at today's broadcast, Hamilton explains, "There's more enthusiasm. You're more animated. There's more competition, and the revenue is unbelievable. And you better do it to the tune of a lot of people liking it. You have a lot of bosses—the sponsors, the ball club, and the audience. You have a lot of different trends of thought you have to please day after day. Yesterday, all they wanted was 'Ball One, Strike Two!' That's why you need the total interest in sports and why you constantly do your homework so that you know everything that's going on in sports."

Another demand of the sportscaster is the interview, those five-minute spots usually before a game with a player, manager, or coach. Hamilton emphasizes that the good interviewer must be a good listener. He insists that he himself know as much as possible about the person to be interviewed. "Interviews can be tough," he says. "Let's say you have a Latin player who doesn't speak English well and isn't too happy about being on the show in the first place. On the other hand, if you don't have him on, he'll think you're ignoring him."

He continues, "Some players do not communicate well and are tough to interview. It's like pulling teeth. That five minutes can seem like fifty. You find yourself talking more than you'd like and more than you should, and you come out with a bad spot because the interviewer did all the talking. The secret is that when you're doing an interview, you want to hear about somebody and from somebody, not yourself."

Hamilton also emphasizes that anyone in sportscasting must get to know the team he is covering. "You've got to have a great association with the ballplayers and the fans. Some people get tired of it, and it shows in their work. They come to the ball park unprepared, and they don't get down on the field and associate with the managers and players. I don't say that you have to do it everyday, but you've got to do it five times a week—especially when a new team comes to town for a series. You've got to get down onto the field and talk to the opposing manager, find out who is hurt, who is playing well or not playing well. You've got to be ready when the game starts. You can't do it after the national anthem. A lot of people who watch the game don't realize that you've done that bookwork, that you've interviewed trainers, players, coaches, and broadcasters and newspaper people from visiting teams. You have to find out all you can, so that if there's any change in the game you know why. That's how you build credibility."

There is also an instinct unique to the sportscaster—the ability to anticipate the next bit of action. Hamilton says that "the biggest thrill" in his broadcasting duties is "setting up the fan for something that's going to happen and have the opposing manager do it."

He explains, "You can't know all the possibilities because there are too many players involved, but you must know the key ones. You have to be able to do things like sniffing out a suicide squeeze. You have to know if the bunter at the plate is good enough to handle the suicide, and if the guy at third is fast enough to score on it. Then when you call a suicide and they try it, the fan at home feels that you know what you're talking about."

Hamilton and other competent sports broadcasters always have much respect for the viewers. As Hamilton says, "Baseball people are the most knowledgeable of all sports fans. Most baseball fans who are with you constantly know as much about the game as you do. You can't fool the baseball fan. You may fool the football fan, because there are so many bodies out there on the field that the average fan doesn't know who made the mistake. But in baseball the players are highly visible. The fans can see everything. So you must respect the fans."

And undoubtedly the potential sportscaster needs self-confidence. Without a moment's hesitation, Milo Hamilton admits it's needed. It all emerges when he says, "Yeah, I think I'm the best. Well, I may not be the best, but 'I'm in the top two!' "

Chapter XIV

DAN MILLER—"I WAS ALWAYS AROUND PEOPLE WHO ALWAYS TAUGHT YOU TO THINK."

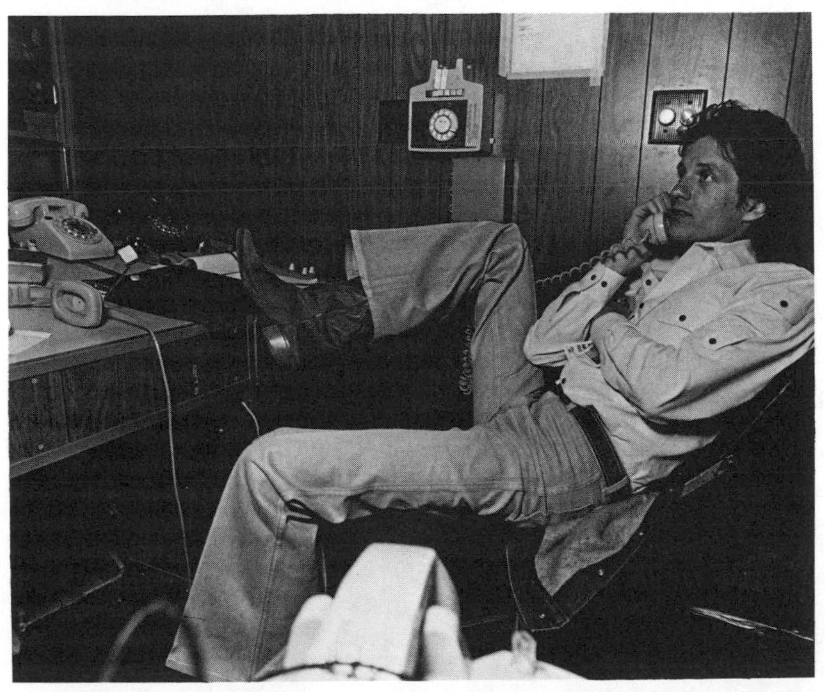

Photo Courtesy of Iowa Broadcasting Network

Dan Miller was an Iowa teenager in the 1960's—that decade that witnessed a surging revolution of the youth in this nation.

There were personal issues—the right to dress as one chooses, the right to wear one's hair to the shoulders, the right to freedom of expression, the right to live one's own life-style.

There were national issues—the struggle for equal rights for blacks, the desegregation of the South, the opposition to the Vietnam War, the lowering of the voting age to eighteen, the movement for the Equal Rights Amendment.

Yes, Dan Miller was only a teenager in Dowling High School, a Catholic school in Des Moines, Iowa—what one might call the heartland of the Bible Belt, the birthplace of conservatism. And Dan Miller probably recognized that he, too, must play a role in the youth movement.

That role would be easy for him to discover—and a natural. As Miller says, "When I was making a decision for a career, there was a history, a legacy, to live up to. My father had been a newsman and was well respected in the journalism business. When that exists, one is tempted at least to explore it by nature."

Explore Dan Miller did while in high school. His father, Marty Miller, was the bureau chief for International News Service (now UPI) in Des Moines, and Dan often visited him at work, observing and admiring his labors in supervising the news gathering and writing processes.

Dan explored every avenue, even talking his way into the public school system just to operate the cameras for the school district's closed circuit television station, KDPS-TV. Less than ten years later that insatiable appetite to be a communicator would lead Dan Miller to his present position as executive producer of public affairs at the studios of the Iowa Public Broadcasting Network.

But there was an additional dimension to Dan Miller's early life that influenced him so much toward a journalism career, and it is evident in Miller's recollection of his father's words, "I can always teach someone how to write, but I can't teach them how to think."

To think and to write. Those two powerful realities are the essence of Dan Miller's background and of his every effort today. He again recalls his youth as he says, "My family was oriented toward classical education. When I asked why I should take four years of Latin, my parents simply told me that Latin would teach me to think and to write."

To think! Miller reminisces again, "I was lucky. I was always around people who always taught you to think because they would sit around and spar all the time. It teaches you to be Socratic. I had the Socratic philosophy throughout my schooling." Then, reflecting upon his work today, he notes, "There are a lot of people who can do, but that's only part of the job. Our job is always to think and to add as many sorts of facts and perspectives and analyses as we can in telling a story."

To write! Recalling his family, Miller explains, "I came from a family where good writing was respected. There would be dinner-table arguments and discussions about language. Anyone can write that news lead—the who, what, when, where, why, and how. But you must write well if you want to pull the reader or viewer further into a story. I always read a magazine like *Newsweek* to find examples of what I think is great magazine writing, especially in the *Newsweek* essays or boxed stories. I save them and pass them around to others."

Thus, Dan Miller talked to a good number of journalists and others he respected in deciding what he would study in college. He explains, "There was a strong body of advice not to major in journalism. They advised me not to narrow myself to that. And I kept remembering my father's words that it was so important to learn to think."

It was such attitudes as these that led Miller to transfer from Iowa State University after two quarters of study to George Washington University, in Washington, D.C., to pursue his career studies. Once there, he spent six years earning a degree in political science while working part-time for Senator Harold Hughes of Iowa.

"I ran an elevator," Miller says with a smile. "But working in Washington is working in a laboratory. I had a great education both at George Washington University and on Capitol Hill. I read the Washington *Post*. I see it as a top newspaper because it is a writing newspaper as opposed to a reporting newspaper. And that's what gets them into trouble at times. But I still would rather see a flourish for the written word in print than just hard facts. If you want just hard facts, you might as well read broadcast wire copy."

The job of operating an elevator did not last too long for an ambitious student like Miller, who recounts, "After six months I talked my way to working in Senator Hughes' office. I learned to operate automatic typewriters and computers. I kept learning as much as I could so I could do more. It's impossible not to learn in Washington. If you want to learn there, you can have a field day."

Miller's experience in Hughes' office soon led him close to the work of the senator's press secretary. He explains, "I worked with the press secretary and I just watched and asked a lot of questions. By doing that, one can develop a good repertoire of skills to do the job yourself. My assumption was always that I was working for reporters, and I tried to get the answer to any question a reporter needed. I served as a liaison just by putting people together."

Then came a one-year stint as press secretary to Iowa's Senator John Culver. Only a year, simply because Dan Miller was itching "to fulfill a long-standing goal to be on the side of the journalists."

That opportunity arrived for Miller in his mid-twenties—an invitation to return to Des Moines to produce a pilot public affairs program called "Farm Digest" for IPBN, the public TV network. He recalls that first assignment in 1975. "They had about four 'Farm Digest' programs during the summer, and I looked at them. They were essentially service-oriented—weather forecasts, farm markets, and a few farm issues.

"But I looked at what Iowans were already getting and I knew they had those services in their newspapers and magazines and on radio. I felt that if I were going to be charged with a show that people would watch, I ought to give them something they couldn't get anywhere else. I knew the state newspaper, the Des Moines *Register,* was doing a very good job out of its Washington bureau on agricultural issues, so I said to myself that Iowans ought to have another place to go if they want issues in agriculture."

And that is just what executive producer Dan Miller did. The first shows under his supervision focused upon grain scandals. Miller says, "It grew from there. We changed the name to "Market to Market," and it grew from forty shows a year for Iowans to fifty-two shows a year for a Midwestern audience."

Today, "Market to Market" is a weekly national offering on 100 television stations in thirty states.

Two years passed, and Dan Miller was presented with another of IPBN's programs, "Iowa Press," a program very much like "Meet the Press" that dealt with state government and politics. But Miller disliked the format, believing that viewers needed some background. He said to others, "I don't care who the guest is, but let's get out into the field and see what the people think the problems are."

Thus, Miller changed the "Iowa Press" format by using part of the program for background. He sent his producers/writers out to tape and record what the people were saying. "I laid a foundation for the viewers for our guests, such as Iowa's Senator Roger Jepsen. I'd show how he switched his vote on the sale of AWACS planes and how he led the fight in the Senate to drop dairy price supports; but I'd also show his work on soil conservation, which is largely unrecognized."

Such a change was not easy, because it cut down the time for the regular panelists and the guest, but Miller cites David Brinkley's remarks in the *Washington Journalism Review*. "Brinkley said the problem with talk shows was that they did not provide viewers with sufficient background and that questions were tossed out and wildly discussed. Reading Brinkley's words was the first time I saw justification for what we were doing."

In addition to the two regular shows, Miller directs the production of all special news and public affairs programs such as the visit to Iowa of Pope John Paul II and the Republican Presidential candidates' debates. He recalls, "We have produced special shows on topics such as Iowa's correctional institutions, addressing the whole subject of prisons and the often misleading notion that people are actually rehabilitated."

As executive producer, Miller has a staff of only six producer/writers, who complete most of the segments for the two regular shows and all specials. The public affairs department of IPBN must share camera crews with other departments, so Miller sometimes hires free-lance reporting crews. But the IPBN staff still travels 60,000 miles a year for the "Market to Market" show.

That small group under Miller is responsible for an average of sixty hours of television programming a year. In addition to supervising the overall production of the shows, the young executive producer must also be involved in money-raising projects, the assignment of stories, and the reading and editing of copy.

But it is the task of producing the programs that excites Miller in his every hour—those extra hours. Or, as he would say, "those extra ten minutes to go below the surface."

And it is probing beyond the superficial that probably keeps Miller in public television news. "We have more time than a commercial station. I think we in public television have an obligation to go further. We have an obligation to work harder."

Miller believes public television must go below the surface of a story because "we hold a dual trust in public TV. We hold the public's trust as journalists and I hold that sacred. But we also do it with the public's money. Most reporters have to avoid conflicts, but we also have to avoid the appearance of conflicts because of that dual trust. I take that very seriously."

"Public TV is public," Miller emphasizes. "It's not just semantic. There are many views of what we should do. We must bring Iowans to Iowans. Part of it is a mandate to bring people whose decisions affect the viewers' lives face to face with them. It's our job to put these people before thoughtful and skilled journalists with appropriate backgrounding. It's our job to let the surrogates of the public or the journalists have a shot at these people."

He adds, "I see public affairs as a responsibility to take important stories that could have or do have impact upon the viewers and treat those stories in depth. To go further and be broader and put perspective upon them. We need to recognize that fine line between perspective and opinion, to show the history and to tell whether the events today mirror that history, because we all learn from that."

Miller has advice for the student exploring a broadcast journalism career. He does not favor a journalism major, saying, "Why go to college to learn to edit video tape? You can stand in a corner and watch and then ask them what they did. If you pay for an education, you really need to be taught—how to think!"

"Start right now," advises Miller, "by working on the school newspaper. Be a station rat—you know, someone who hangs around a TV station to learn everything there is to know."

And then, Miller says, "Be curious. If you're curious, you'll learn how to get the information. Pick up a rock, if for no other reason than to see what's on the other side. You'll learn to write better that way. Be critical of your own work and that of others. Be critical of what you hear."

The time element? Miller quickly warns, "You've got to react immediately to all current events. You have to love to be curious and interested in what's going on all the time. If you want a 9-to-5 job, you can't work for me. It's a 24-hour job."

Perhaps Miller best summarizes his whole attitude as he advises the prospective journalist, "It is a commitment to get it all out there—you are a conduit of information. A funnel. Get the news out to them and give that news some meaning."

Chapter XV

AN EDUCATOR LOOKS AT BROADCASTING

It is evident from interviews with both new and veteran journalism broadcasters that there is no single route to a television broadcasting career. Some have earned journalism degrees; others have degrees in political science, economics, English, and other subjects. However, few can enter news broadcasting without a degree of some kind.

Thus, many high school students seeking a journalism career discover a perplexing problem in deciding what college to attend and what area of study to pursue. Questions arise: Do I want to work on a newspaper first? Need I study journalism? Would I be satisfied with the limitations of TV news reporting? Can I survive the "show-biz" aspect?

Certainly, most television news executives recommend at least some journalism training, whether it be specifically in electronic broadcasting or in general journalism. Many would advise the prospective student to study carefully the journalism program and the degree requirements. One of the professional journalism organizations that is concerned with all areas of journalistic concentration is the Association for Education in Journalism (AEJ), a national organization of schools of journalism in colleges and universities.

The current president of AEJ is Dr. Kenneth Starck, director of the School of Journalism and Mass Communications at the University of Iowa. Recently the author visited informally with Starck, seeking his views concerning television news as broadcast today, the state of journalism education today, and television news as he sees it for tomorrow.

AUTHOR: Dr. Starck, let's start out by looking at those three network news broadcasts. What do you see?

STARCK: I think they serve pretty much as a headline service. In many respects, it seems to me they don't qualify as journalism, particularly the news shows themselves. They seem to have taken off as a "Star Wars" quality with all the electronically computerized graphics. The stress primarily is on *what* is happening in a service way rather than an attempt to dig beneath the headlines to provide a thoughtful, contextual interpretive account of the events.

AUTHOR: Are there exceptions?

STARCK: Yes, the "McNeil/Lehrer Report" on public television is the most notable exception, I suppose. Another is "Nightline" on ABC-TV. Now the other networks are toying with the possibility of some sort of an in-depth, journalistic account of the day, late in the day.

AUTHOR: Some critics look upon the newscasters as "show-biz" personalities.

STARCK: Basically I think the people who perform on TV news broadcasts are just that—they perform! You get a few who are established journalists, who have won their journalistic credentials through successful careers of actually studying something and reporting about it in an informed way; but so much of the actual journalism is done by journalists who never come on camera. They are the ones who have the background to understand events and to present them in an articulate fashion. Those on the screen—the nicely dressed and nicely manicured folks—they just read!

In fact, "60 Minutes" at one time may have had some good journalism. I'm not so sure anymore, with some of the accounts that have come out about how Mike Wallace and others have worked to develop those segments. They do topics of current interest and involve some pretty heavy investigation, but it seems as though they are so concerned with turning it into a dramatic production that there have to be good guys, bad guys, beginnings and endings, and a show-biz aspect. It has gotten out of hand.

AUTHOR: Certainly you must view those television newscasts, if only from academic curiosity.

STARCK: I like the six o'clock news from a standpoint of entertainment. It's a good show. They've got the 30-minute production. It's well thought out. They have some information that is useful in a visceral kind of way. You know, you've got the latest of what's going on in the Falkland Islands or in the United Nations or what the Chicago Cubs have done earlier in the day. Even though much of the news tends to be grim and presents a bleak outlook, usually a couple of fluffy stories at the end suggest that people aren't as bad as they're made out to be. It's entertaining and informative to some extent. If you're not careful, it can be addictive. But it doesn't really provide much.

AUTHOR: But don't we need that newscast to know what is happening?

STARCK: I've often thought about that. Would it be better not to try to provide any information about an event than to distort it to the point where, in essence, you provide misinformation? That is very disturbing. That's lousy journalism. It fulfills a kind of entertainment value. I think it's irresponsible. And I know when I say that I'm indicting about every TV operation in the country. The same thing happens in Waterloo and Cedar Rapids. To a great extent, stations have entertainment values. They get sponsors. They get into a lot of electronic graphics and gadgetry.

They buy new equipment to bring forth a live news event—tune in at 6:10 and get live coverage of a picnic or something that is not significant.

What about a live report of a city council meeting or a school board meeting? In fact, this is where I see television having an impact, particularly through cable, in telecommunications. Just raw data. No commentary. It isn't good quality, but it's there if you want it. I can see it at special school board meetings when important policy is being considered.

It may be that somewhere down the line newspapers are going to be headline services. Newspapers are going to provide a summary of school board meetings. Cable will give you all of it. Cable enables you to do it because of the minimal cost. But traditional TV looks for the biggest audience for the biggest profit.

AUTHOR: Doesn't that bring up the question as to what is news?

STARCK: That opens up avenues. News, it seems, has to be defined in such a way that it has audience appeal. It will have to have sufficient appeal that they'll want to pay for it. I think economics determines news more than we think it does. Maybe I'm naive or overly optimistic. I don't think people are going to be any less informed than they have been. They will probably be informed in different ways by different media. I think that, with the proliferation of cable and changes in the whole telecommunications area, you'll find some very good journalism practiced on TV. Certain channels and organizations such as Cable News Network will provide opportunities for the development of programs that people will pay attention to and that will provide the raw material to help keep people informed. I think we'll continue to have newspapers, but in a different fashion than what we presently have. We may not rely upon them quite as exclusively as we have in the past, but there are some things you can do in print that are not possible on TV, although the opposite holds also in terms of the visual.

AUTHOR: Will we see that newspaper of the future in the form of a printout emerging from the TV box in the living room?

STARCK: I think we're a long way from being able to provide a printout of news accounts electronically. It's possible to do it, but it's extremely expensive.

AUTHOR: I often fear we will be less informed in the future.

STARCK: I don't think we're going to have a less informed citizenry. One thing that does concern me about this is it may eventually work toward developing more differences among people. Regardless of the quality of media, you're always going to have some people who are misinformed or poorly informed. They don't pay attention to the media. This may be tied to economics or geography, but by the same token, you are going to have some people who are informed. As the media get better and as more information is available, people are going to pick up more of it and pay more attention to it.

What we might find in this country is a greater split between the informational "haves" and "have-nots"—not unlike what one finds globally between the "have" and "have-not" countries. Even though some steps have been taken to reduce this gap, I don't know that they've been effective globally. I don't know if they'll be effective in this country unless there is some kind of policy that permits those who don't "have" to gain access to information.

AUTHOR: Are we talking about marketing the news?

STARCK: Yes. I was at an American Press Institute Meeting a couple of years ago and some people came in from Chicago, Washington, and New York newspapers. They were talking about marketing—"down markets" and "up markets." These were editors! Editors talking about finding out what readers wanted and what income readers had so that the medium would have clout with advertisers, so they could say to advertisers, "We've got 5,000 readers out there with $50,000 or more a year." Fine! But what about the readers at the other end of the spectrum? Or the potential readers? They're probably not readers because the product ultimately is being packaged. The news content is being determined by those who are able to buy the product. The 15,000 people living in the ghetto don't represent a viable market for the advertisers, so as a result we write them off as a mass medium. We really are not much interested in carrying them as subscribers, which means that in our definition of news what goes on is probably not news unless it has an impact upon those 5,000 people.

AUTHOR: Once again we come to the question of how people are to be informed of news events.

STARCK: Throw out TV. And newspapers and radio. Would people be totally uninformed about what is going on in their sphere of interest and influence? I don't know the answer. I am certain there would be some level of informedness by virtue of attendance at council meetings, school board meetings, and so on. They learn through calls, conversations, conventions. They stay informed in that way on matters that concern them.

AUTHOR: That gives a prospective journalist a good deal to think about. Let's turn now to the journalism student. Just how do you think journalism students should prepare themselves in college?

STARCK: If they are really interested in journalism—and that means essentially, I like to think, an intellectual enterprise—they should dig into academically substantive areas in their academic program. They should be basically interested in a liberal arts preparation, with their course work determined by their interests. If they're interested in public affairs, then get into political science courses. I think that a little bit of production and technique goes a long, long way. I think a good school of journalism program will focus upon substantive areas, will permit the student to get substance that they can't get in their own program by enrolling in other departments, and will provide some technical instruction, some production.

But keep it in perspective! What does that mean? I'm not altogether sure. I usually approach it from the standpoint of trying to get a balance between theory and practice, though I don't like setting off those two notions in a bipolar way because it seems to me that they are really intertwined, even when you're teaching a fairly practical course such as introduction to journalistic writing. This is a whole new ball game for them. I think you try to stress the principles, information, ideas that are going to be long lasting for the student. It isn't just a matter of mastering the inverted pyramid, but thinking in a way about the information, about the story or the idea that they are working with. Thinking in the terms of the consequences of it. They should come away with some of those things that have a lot more lasting value for the student than just the technique, for which they might be unprepared anyway, with the constant changes in technology today.

AUTHOR: That seems to be a problem that every school of journalism faces—finding a fair and profitable balance for the student between theory and practice.

STARCK: I agree. How do you set up an environment? How do you conduct a classroom where you encourage that endeavor, where you bring out the thinking on the part of students? We get into a debate around here about the very basics of journalism instruction and of education overall. Not long ago we had a big debate in essence about whether we should accept a $25,000 gift to buy several video display terminals or whether we can continue to teach what we want to in the editing and copy process without the equipment. Would the equipment possibly limit us in our instruction? In some way would it ultimately determine the instructional content? We do not want what goes on in the classroom to be determined by machinery. Well, we resolved it by buying the machinery and providing the students a little hands-on experience. But let's make sure the instructors keep that in perspective, because though students can master the machinery that is here, that doesn't enable them to go out and operate in a different system. But if we stress the principles sufficiently, and if we stress the need for students to be aware they must continually learn to adapt and approach their careers and their lives in this way, then they're going to be able to cope better with different systems and with the changes that occur in the profession.

AUTHOR: To provide that substantive approach, I note that many schools of journalism are requiring a double major.

STARCK: The better journalism programs are set up to bring that about. There are a number of us who require a second major. A student can't get through this major at the University of Iowa without doing one of two things—getting a major in another area such as political science, which is fairly popular among the serious students, or developing a specific area of concentration on the basis of his or her own interests and aspirations.

AUTHOR: What is important for the student to remember in this double-major approach?

STARCK: It partly boils down to the fact that if you teach or stress technique, you help students learn how to do something, how to write a story. But if they don't know what they're writing about, I think you wind up with shoddy journalism.

AUTHOR: Doesn't that bring us to the problem of knowing how to communicate to others? The reporter preparing his news story for either a TV broadcast or a newspaper?

STARCK: There has to be some kind of intervention that stresses the communication aspect of writing—thinking about something in a communicative sense. The simple idea of being aware that you might be addressing an audience that is not as well informed about a subject as you are. This happens with the scientific disciplines. You get a bunch of clinical psychologists together and they can understand each other, but to you or me it's just jargon. You must build up some awareness of what's involved in the communication process, whether it be interpersonal or mass communication. I think the bright student might catch on to some of these things. But instructionally, I think we can bring these things to the attention of students so they can be aware of how they themselves can behave, whatever their objectives happen to be.

AUTHOR: And the audience?

STARCK: When you talk about awareness, that opens up some avenues. On one hand, you can talk about the mass consumer, who is a kind of faceless lay person. If you have to write in that way as a journalist, you're not only one who is informed but also in a translator role. By the same token, there is a kind of journalism that appeals to certain segments of the audience. In fact, there is one phenomenon in mass media today that explains to some extent the demise of metropolitan newspapers: they simply are not appealing to the wide spectrum of the audience as they once did. So you have kind of a "demassification" of communication where the member of the audience knows what he wants and doesn't want to be bothered with anything else.

AUTHOR: Let's move now into some of the characteristics or traits you believe a television broadcast journalist should possess.

STARCK: First, I think any broadcast journalist would be a lot better off in learning about the work-a-day world of journalism by working on a newspaper rather than starting at a TV station, because I think that many broadcast operations do not practice journalism. They practice a sort of pseudo-journalism. Journalism is not picking up the Des Moines *Register* and reading from it. Journalism is putting together the Des Moines *Register,* and there aren't many stations that will make a firm commitment, financially and otherwise, to getting their staff really to practice journalism. I would prefer a person who has had newspaper experience. I would certainly want someone who has gone to a journalism school and has had the opportunity to do some of the things associated with newspaper journalism—by that I mean have a mastery of collecting information and writing it. It is amazing how many students come in and say, "I'm not much interested in writing, that's why I want to go into broadcast journalism."

As consumers of mass media, they get an incorrect view of what it takes to be a journalist. It's an illusion that the media themselves create.

AUTHOR: What about some personal traits?

STARCK: One has to take the initiative, be resourceful, and certainly have a sense of social responsibility. I think one should have an outgoing personality. One should be reasonably intelligent, well read, and well informed generally.

AUTHOR: You said "a sense of responsibility." I think we need to elaborate upon that point.

STARCK: Social responsibility is being aware of the role the mass media play in society. It is the only business protected by the First Amendment, which suggests that it is more than a matter of making money.

It's a matter of the mass media having a particular responsibility that goes along with the protection of the First Amendment. They have a responsibility to help bring about an informed electorate. As a result, one has to be aware of serving two masters. On one hand you are serving the person or organization that's paying your salary. On the other hand you're serving the larger community, or society. Sometimes those interests clash with each other, and you face the problem of resolving the clashes; in an extreme situation, you might risk your job because of your personal sense of social responsibility. There are certain things that seem inappropriate to you, and if a situation reaches that point, either you quit or are fired or you acquiesce to something else.

AUTHOR: But just how does one determine that sense of responsibility?

STARCK: John Merrill, author of *The Imperative of Freedom,* says that in the final analysis ethics and responsibility are determined by the individual in a kind of personal, conscionable way and that, in essence, there is no social responsibility. Rather, there is a collective responsibility—and you yourself determine what the standards are. Well, I think that to some extent those standards are determined socially, so you have a kind of interplay between what your individual conscience tells you and what your social conscience tells you. I think that is how you try to resolve these things.

AUTHOR: A final note, Dr. Starck. You have a record enrollment here in the school, as is true at many colleges. Will there be jobs for all the graduates?

STARCK: Yes, but it isn't just in journalism. Remember that many schools have added the total field of mass communication. The boom in enrollment is in that field, not journalism. This will open up many new worlds of opportunities in business, industry, and professions. We haven't even identified all of them yet, but the opportunities are there.

Chapter XVI

WHAT THEY WANT FROM YOU

When the college-trained broadcast journalist applies for that first job, he or she will undoubtedly be interviewed by the news director of the radio or television station. That executive will be looking at the applicant's education, journalistic skills, and attitudes. Michael Bille, news director at WQAD-TV in Moline, Illinois, conducts such interviews. At only thirty-two years of age, he has earned his college degree in journalism, worked as a reporter and weatherman on a small television station in Minnesota, and then moved to his present station—first as a reporter, then an anchorman, and finally news director.

In the following interview, Bille concentrates upon those characteristics and skills that make a journalism broadcaster successful. It can well serve as an informal checklist for the aspiring journalist.

AUTHOR: Before we start talking about reporters, will you please briefly summarize your own career. I think it is important because of the fast rise you have experienced.

BILLE: Well, I was born and raised in Minneapolis. I attended the University of Minnesota, where I earned a degree in journalism. While there I worked on the university newspaper, the *Minnesota Daily,* and served as assignment editor, opinion writer, and book reviewer. However, I did take one year off from college before I was graduated and went to Washington, D.C., where I wrote and edited a monthly newspaper for the National Welfare Rights Association. It was a terrific experience because I learned the inner workings of Capitol Hill and how the media work in Washington. It was a basic left-wing organization—disorganized but an exciting time for liberal politics.

AUTHOR: What was your first job after graduation?

BILLE: First I went to England for about a year with my wife, who had been awarded a scholarship to study over there. I didn't have a job. It was 1973, during the coalminers' strike, and it was a tough time in England for everyone. I got hooked on radio plays. I still have the ones I wrote, and someday I hope to work more on them. But being in England gave me a chance to practice and experiment with my writing abilities in something other than nonfiction. Within a year I came back and was fortunate to get a job with an ABC affiliate television station in Austin, Minnesota. I went there as a weatherman, and it was also my job to write one news story a day. I was there eighteen months and then came here. When I started I was a reporter and served as weekend anchor and producer for a couple of years. The Des Moines *Register and Tribune* bought the station, brought in new executives, and eventually I was given this news director's position.

AUTHOR: Briefly describe your duties as news director.

BILLE: In simple terms, I'm responsible for everything that happens in the news department. Above all, it's my job to maintain journalistic integrity—to make sure that we have the facts in our stories so we don't get sued. I also do all the hiring and firing. I make up the news department budget for the year, and I get involved in the community as much as I can. I've also worked in our professional organizations and have done a lot with the camera in courtroom procedures in Iowa.

AUTHOR: You say you do the hiring. What are the qualities you want in a reporter?

BILLE: Most of our staff has had experience in the smaller markets, so I'm not looking so much at their college degrees. A journalism degree never meant much to me. I think a college degree is very important, however, because college gives you the opportunity to mature and to experience different ways of life. It's the broad-based education you get in college more than the degree itself. It's a love for politics and issues and the desire to write that I think is important to get out of college.

AUTHOR: What procedures do you use in the hiring of reporters?

BILLE: I look at the journalistic skills first. I view videotape samples of a reporter's work. I want to see the writing samples. But it is the interview that I really need.

AUTHOR: You must be looking for something special in that interview.

BILLE: Yes, I'm looking for a basic inquisitive nature—people who have been involved in a lot of different things in life. I guess it's that category called hobbies. I don't mean reading or woodcarving, but I want to know if they served on the high school or college newspaper. I want to know if they belonged to Young Republicans—just want to know if they were involved in the same kinds of things on the same kind of level that they will experience when they become professionals. I think if a person was involved with politics in college, that person will have a head start in the professional world in covering politics because it will be second nature to them. If you didn't care about politics in college, I think you'll have a hard time covering the first day on the job here as a reporter.

AUTHOR: Why is that so important?

BILLE: You've got to be interested in a lot of things in this reporting life. Every day is different. One day you'll be working on a medical story and the next day on politics. Still another day you may be working on a human interest story. You're doing an urban story one day and a rural story the next day. You don't have to want to be a farmer, but you should be interested in that aspect of life, too. I frown upon those who say they don't want to do an agricultural story. Maybe some day they'll be in a market where they can be selective, but when you start out in a market such as ours, I would want reporters interested in everything here.

AUTHOR: You said you look at videotape samples of their work. Again, precisely what are you looking for in those?

BILLE: I look at the totality of the story and ask if they got to the root of the issue. Do I, as a viewer who may have no background on the subject, get enough explanation so that the problem is set up, all aspects laid out—you know, almost like a short story, with a beginning, a middle, and a conclusion. Did the reporter address all the issues? Were the pros and cons there? Did the reporter wrap it up in such a way that I understood what it was all about?

Then I look at technique. Do they speak in a way that's clear, pleasant, understandable? Is the writing clear, understandable? I look at the way the reporter produced the story. The film technique, the use of natural sound to make the story more interesting. But I don't think the technique is as important as what I mentioned before. I just hired a newspaper reporter recently. He didn't know the technical aspect of television at all, but he had the inquisitive nature. He knew the town, the issues. He is a journalist. He's been involved in things all his life. He was the kind of person I look for.

AUTHOR: What about the show-biz factor?

BILLE: There's a lot of show business in broadcast journalism, and I think we've taken a bad rap for that. Look, if someone goes to a movie, he expects some critic to have told him beforehand about that movie. He expects the critic to say that a certain actor did not portray a character well, that he wasn't believable. Well, TV is that way, too. If you don't talk in a pleasant way that people enjoy hearing, and if you don't use some theatrical techniques to make people perk up their ears, then you're not doing your job as a TV reporter. If you feel you're compromising your journalistic skills, then you probably should write for a newspaper. But don't forget that a few theatrical skills are used in newspaper writing, too.

AUTHOR: Does that mean you would fire someone for that reason?

BILLE: Yes! But there are several reasons why I fire a reporter or talent. First, some make too many mistakes. They just aren't good journalists. I have to fire them because they do not pay attention to the detail that makes the difference between a storyteller and a journalist. They can put me in the position of having to ward off possible lawsuits. And that has happened. After working with these people time after time, you can afford just so much. I have to tell them either that they don't have it or that they will have to go somewhere else to learn.

Second, there are attitude problems with some. As reporters they didn't like to do certain kinds of stories. It showed in their demeanor with people they were interviewing, and it showed in the way they presented their stories. How is the public watching the 10 o'clock news going to be interested in the story if the reporter isn't?

Third, I've fired air talent because they simply weren't coming across as strongly as the competition. I let go a man who had been here ten years as a weatherman. I had researched and discovered that he was less popular than the weathermen on the other two television stations who had been here less time than he. I've had to fire some anchorpersons. They had good voices, good looks, but they weren't good communicators. You can say I've also let people go simply for theatrical reasons. It sounds cold and ruthless, but the public is not so naive that they don't know about this. It really is a cut-throat business. It is a business that relies upon numbers. We are mass media. It's no different than a new toothpaste that goes on the market; if it can't capture a certain share of the market, the company loses money and can't afford to make it anymore. What good is the best journalist in the world if nobody is watching?

AUTHOR: Do you feel any pressure with that?

BILLE: Yes, I feel the pressure. But this station is owned by a great newspaper, and they have a tremendous commitment to journalism. I've never been told to get numbers. But I have been told to put out a quality journalism product—something that is responsive to our community and to our profession. The company believes that if we do that, the numbers will follow. So far, they've been right.

AUTHOR: Take the reporter. He's both a journalist and perhaps a performer in terms of theatrics. That could mean there is some competition among the reporters.

BILLE: I hope so! I hope our reporters compete. I'd like to see more of it. I like to see that aggressive attitude that everyone wants the lead story on the newscast. It's a sense of pride for them. I want to see them clambering all over each other for that lead story.

AUTHOR: You told me that you have experienced a strong sense of pride when one of your reporters moves on to a bigger television market. It means that you are doing something right here. And I sense that you feel a strong commitment to broadcast journalism. How would you want to describe that commitment to prospective broadcast journalists?

BILLE: I can't think of another profession in which I could meet so many people with so many backgrounds, where I could deal with so many issues that affect so many people day in and day out. The variety of the job just for personal satisfaction is incredible. I guess I'm one of those people who have always been interested in politics and in the community. Reporting is the best way to do it. You're going right to the heart of it. While others must watch it on television or read it in a newspaper, I get to talk with that politician, that Congressman. I can talk to the guy who has been charged with a crime. It's fascinating to hear a lot of news first.

As for ego, it's just an incredibly strong thing to know that 50,000 to 60,000 households out there are watching the thing you worked on all day long.

If there is someone out there who feels he or she would like all of this, you tell them to start now!

Chapter XVII

WHICH JOURNALISM SCHOOL TO CHOOSE?

The toughest task for many prospective broadcast journalists is to select a journalism program in which to prepare themselves for careers. As so candidly demonstrated in the interviews in previous chapters, there are a number of collegiate routes—journalism majors, English majors, political science majors, double majors, and so on.

Sometimes it seems that no technical training in college is required for those seeking television newscasting careers, and for a select few, that is true. However, the aspiring journalist needs to remember that broadcasting, particularly television, is a relatively young industry, and it is likely to expand greatly both in technology and opportunities. Thus, as it grows, broadcast journalism will demand expanded skills and knowledge.

Also, the prospective broadcast journalist must recognize that it is rare for a person to enter a television newsroom without a college degree, whether in journalism or not. It is virtually impossible to obtain even the lowest position in any broadcast journalism operation without that "sacred sheepskin." And, as broadcast journalism expands, this educational demand will strengthen. Already, there is evidence of a good number of broadcast journalists with credentials that include both bachelor's and master's degrees in mass communications.

Because of the tremendous growth of television, and especially with the emergence of cable television, hundreds of two-year community colleges and small four-year liberal arts colleges across the nation are offering journalism and mass communications programs. This quickly creates a dilemma for prospective journalists, the chief question being which is the best journalism program?

Though the arguments are many, there are two major criteria for students to consider in deciding which journalism school would be best for them. First, study carefully the facilities, faculty, and curriculum of the program. Second, determine if the school is accredited by the American Council on Education in Journalism and Mass Communications.

This chapter will explore these two criteria in some detail.

Facilities, Faculty, and Curriculum

In examining a journalism program in a college, prospective students should visit the campus and tour the facilities. Is there a daily newspaper? Or weekly? Is there a college radio station? Is there a closed circuit or cable college television station? Do these facilities provide adequate opportunities for students to gain practical experience? Such questions can be answered only by personal visits to the college and conversations with both faculty members and students. On far too many campuses only the barest of facilities exist. In far too many situations only a minimum number of opportunities are offered to students to be involved in the total operation of the facilities.

Next, prospective students need to study carefully the faculty involved in journalism training. Are there only two or three instructors teaching the journalism courses? What are their credentials? Is there a heavy reliance upon part-time instructors from local high schools, businesses, and media? Is there a good deal of dovetailing of courses and instructors among various departments? Too often the program is not staffed by educators with extensive training and experience in journalism.

Third, and perhaps most important, is the curriculum. Some colleges offer a minimum number of courses. Others offer a wide range of electives. Still others offer the double-major programs. The following courses are currently offered by the School of Journalism at the University of Kansas, which awards the bachelor's, master's, and doctoral degrees in journalism. The list is not exhaustive but is offered as a guide to the kinds of courses available. In compiling the list, emphasis has been placed upon broadcast journalism courses.

Introduction to Radio and Television: A study of the history of radio and television; their social, educational, and commercial significance; radio and television as media for communication, entertainment, and dissemination of information. Discussion–laboratory.

Promotional Writing: Basic principles of English and news writing applied to advertising areas of promotion, publicity, and public relations. Case studies provide practice in writing media releases.

Reporting I: The role of the reporter in communicating public intelligence. Basic principles of news writing. Copy to be typewritten.

Introduction to Writing for Radio and Television: A survey course to introduce the skills and cognitive processes involved in the writing of nondramatic radio and television continuity, namely: announcements, short features, talk shows, standard formats, news, and programs for special target audiences. Comparative techniques in writing for the ear vs. the eye.

Laboratory in Radio: A laboratory course designed to give a student experience under station operation conditions. The student will perform in selected staff assignments on the FM station KJHK.

Radio Programming and Operations: A lecture course examining the programming and operations techniques in radio. Students will learn various radio formats, FCC rules and regulations, internal station organization, and staff functions and responsibilities.

Film Production: An introduction to basic principles of double system cinematography, sound recording, and editing.

Basic Television Production: Production-direction theory and operations: preproduction planning, scripting, directing, audio production, video lighting, and camera and switcher operations for various formats. Lecture–laboratory.

Broadcast News I: Reporting, writing, and editing the news for broadcasting. Emphasis on the use of tape recorders, beepers, silent and sound-on-film cameras, and the production methods used in putting the materials on the air.

The Art of Interviewing: Course treats the fascinating techniques and practices of oral interviewing. Skills of gathering and sorting information through the use of investigative reporting, eyewitness accounts, participant observations, and other types of field work procedures are developed. Students work on projects of their own selection while demonstrating the usage and understanding of these tools. Continuous feedback from participants and a close working relationship with the instructor are featured. Role and contribution of oral methodology for journalism, history, and the social sciences are discussed.

Photojournalism I: An introduction to the basic principles of news photography. Course work includes taking, developing, and printing news and feature pictures for the mass media.

Color Television Production: Color television production in the studio. Preproduction and postproduction work, including script writing, talent coordination, and tape editing.

Television Programming: The analysis and planning of television station programming structures and schedules. The role of programming in broadcasting, factors that affect programming, general concepts of station programming, applications of audience research. Public service broadcasting.

The Documentary in Broadcasting and Film: A study of the forms and functions of the documentary; its development and achievements in print, photography, radio, motion pictures, and television; its problems and possibilities as a mode of sociopolitical persuasion in our society.

Writing for Broadcasting and Film: Writing of dramatic and nondramatic scripts for radio, television, and film. Emphasis on problems of format, structure, narration, dialogue, character action, with special attention to creative use of the visual image.

Advanced Media Studies: Broadcasting and Film: The utilization of advanced principles of operation, programming, and production. Through choice of the appropriate course sections the student may participate as an intern at specific area stations or production companies.

Television News Production: Television news production will examine the elements of news production—formats, scripts, sets, talent, and visuals. Students will produce, direct, and crew television newscasts.

Broadcast News II: Advanced study of the preparation and production of the newscast and the news documentary. Presentation of these types on a regularly scheduled basis using the radio and television facilities of the university.

History and Criticism of American Broadcasting: An examination of the history and development of the commercial broadcasting system in the United States; comparison with foreign systems of broadcasting, and an analysis of contemporary criticism and trends.

Station Organization and Operation: Theories of management; study and analysis of the special problems and situations confronting the manager of the broadcasting station in personnel, operations, government relations, programming, and sales.

American Broadcasting: Designed for the journalism student with a serious interest in selected problems, issues, and recent developments in the electronic media's news operations.

Mass Media and the Popular Arts in America: "Mass culture" and its relation to the mass media, written and electronic, and the popular art forms. Examination, through representative examples, of such institutions as books, newspapers, magazines, television, radio, the movies, music, and painting within popular American culture.

Law of Communication: This course is designed to examine the historical background of the concepts of freedom of speech and freedom of the press and the limitations that have been imposed on them by statute and by common law. The case study approach is used, but the emphasis is on the principles and the philosophy that underlie the landmark cases.

Accreditation

Accreditation simply means that a college program is approved by a governing body as meeting a certain set of standards. Accrediting procedures and standards exist in practically every area of academic study in higher education—in medicine, law, education, engineering, English, architecture, journalism, and so on.

For journalism, the accrediting agency is the American Council on Education in Journalism and Mass Communications (ACEJMC). The council has thirty-three members, who include nineteen professionals from such organizations as the American Newspaper Publishers Association, the American Society of Newspaper Editors, the Associated Press Broadcasters Association, the Broadcast Education Association, the National Association of Broadcasters, the Radio-Television News Directors Association, and Women in Communications.

The council also includes eleven educators, all from accredited schools of journalism and mass communications such as those at Columbia University in New York, Boston University, the University of Missouri, Colorado State University, and the University of Southern California. The other three members are from the general public.

The council lists six benefits of accreditation:

1. To provide a service to society: to students, parents, faculty, employers, and the public at large.
2. To insure continued improvement of the quality of instruction through reevaluation at six-year intervals.
3. To provide administrators and faculty with the stimulation of outside viewpoints and to exchange opinions when they differ.
4. To encourage research in journalism and communications.
5. To provide an agency capable of taking action on complaints by students, faculty, and the public.
6. To insure that journalism education, basically professional, continues to encourage both breadth and depth of students' exposure to the liberal arts, the sciences, and the social sciences.

Following is an outline of the ACEJMC's accrediting standards for undergraduate and graduate programs. In the statements that follow, the word "unit" refers to the overall program in journalism or mass communications. The word "program" is used to describe the individual sequences such as advertising, news-editorial, and radio-television broadcasting.

1. UNIT OBJECTIVES AND GUIDELINES
   a. A unit should state its objectives in as concrete a form as possible, including specific objectives for sequences offered, and these objectives should be published in its catalogue and descriptive literature.
   b. A unit should claim to educate students only for those areas of this broad field for which it has competent faculty, adequate library facilities, and appropriate equipment.
   c. A unit should be evaluated in terms of its stated objectives. A small school that claims to prepare students for reporting assignments should be judged on the basis of its claims. A school that claims to offer programs in several of the various phases of journalism and mass communications should be judged on the basis

of its claims. If the objectives prove to be either too limited or too diffuse, the Council may consider the school's program beyond its purview.

d. A unit should require its undergraduate students to achieve a ratio of approximately three-fourths–one-fourth of broad liberal arts and sciences to journalism. As a part of the examining process ACEJMC will also review the quality and depth of the liberal arts and sciences segment of the overall curriculum. Since graduate students are admitted individually and on the basis of several criteria, ACEJMC will examine graduate student credentials separately.

e. A unit should strive for a ratio of 15 to 1 of students to professors in all seminars and laboratory classes.

f. As a guideline, ACEJMC expects that more than 50 percent of the students in a unit seeking accreditation should be in accredited or accreditable programs.

g. No unit will be examined if it does not also request accreditation for at least one program.

2. BACKGROUND EDUCATION

a. Study in journalism and mass communications should be based on a wide and varied background of competent instruction in the liberal arts and sciences.

b. The unit should be located in an institution with a four-year accredited program in liberal arts and sciences.

c. The liberal arts and sciences background of the student should include wide study in such fields as economics, English, ethics, history, languages, literature, philosophy, political science, psychology, sociology, and the sciences, as well as depth in one such field.

d. ACEJMC expects an accredited program to provide the student with basic skills, and also to encourage the motivations of a working professional. The number of credit hours required to provide this vital part of the student's professional education may vary from school to school, but the essence of its fulfillment is considered a requirement of all accredited divisions of journalism and mass communications.

3. PROFESSIONAL COURSES

a. The required professional courses for a program should vary with the objectives of the program or sequence, but all students should be instructed in the basic elements of factual writing, editing, communications law, ethics, and the theory, history, and responsibility of journalism and mass communications.

b. A program should concentrate its professional courses in the last two full-year professional courses below the junior year. The purpose of this standard is to permit the student to acquire a basic background in the liberal arts and sciences.

4. FACULTY
   a. The number of full-time faculty members shall depend upon the scope of sequence offerings. Any program offering, however, should embrace a faculty of sufficient size, academic attainment, professional experience, and diversity to provide realistic instruction, research, and service in the areas of concentration.
   b. A faculty should bring appropriate professional experience and advanced academic preparation to its students in the areas in which instruction is offered. It is recognized that certain courses are enriched more by professional experience on the part of instructors than are others. It is further recognized, however, that there are points of little return in long periods of professional service, just as there is little to commend part- or short-time periods of service in the profession as grist for enlightening professional experience. ACEJMC's increasing emphasis on the student's need for a broad general education should not be interpreted as an abdication of interest in the need to bring the experience and insight of the practitioner into the classroom.
5. FACILITIES
   a. A unit should have facilities adequate for the objectives that it has established.
   b. A unit should have available an adequate collection of library materials in professional journalism and in the social sciences and other areas related to journalism and mass communications.
   c. Special facilities should be available if the unit proposes to prepare personnel for special fields.
6. GRADUATES
   a. The professional performance of graduates should be considered as a major item in the accrediting evaluation.
   b. To offer an acceptable program, a unit should produce at least five graduates a year.
7. RELATIONSHIP WITH THE MEDIA AND WITH PROFESSIONAL ORGANIZATIONS

   A qualified school should assume an obligation to maintain a working relationship with the various media in those areas in which it offers educational programs and should cooperate with professional organizations for the maintenance and improvement of standards.
8. INTERNSHIPS
   a. When academic credit is granted for internships, the institution should validate the learning experience through (1) advanced planning with the employers, (2) monitoring during the intern-

ship, and (3) evaluation of the experience by both the students and the employer.

b. No more than 10 percent of a student's journalism and mass communications credit can be earned through such internships. Work on student media can be defined as an internship.

In summary, it would be folly to claim that one can obtain a good preparation for a broadcast journalism career only at the college with an accredited program. But for many, there is a strong bias for the accredited program simply because of the extensive standards established for facilities, faculty, and curriculum. Many of these accredited programs are available at state-financed colleges and universities, which is a factor in tuition costs, living costs, and transportation.

The prospective broadcast journalist should study carefully the various possibilities by a thorough examination of the program and a visit to the school to determine if its program meets his needs and wants.

APPENDIX A

ACCREDITED COLLEGES AND UNIVERSITIES

The following colleges and universities offer journalism programs accredited by the American Council on Education in Journalism and Mass Communications. Catalogues, brochures, and other forms of information may be obtained by written request. Such information will include requirements for graduation, required and elective courses for a journalism major, description of facilities, scholarship opportunities, faculty listings, and financial costs. Prospective students should obtain information from a number of journalism schools for study purposes in selecting a school to attend.

University of Alabama, University, AL 35486
School of Communication
William H. Melson, dean

The American University, Washington, DC 20016
School of Communication
Frank J. Jordan, dean

Arizona State University, Tempe, AZ 85287
Department of Journalism
ElDean Bennett, chairman

University of Arizona, Tucson, AZ 85721
Department of Journalism
Donald W. Carson, head

Arkansas State University, State University, AR 72467
College of Communications
Robert L. Hoskins, dean

University of Arkansas, Fayetteville, AR 72701
Department of Journalism
Roy Reed, acting chairman

University of Arkansas, Little Rock, AR 72204
Department of Journalism
Leonard A. Granato, chairperson

Ball State University, Muncie, IN 47306
Department of Journalism
Mark Popovich, chairman

Boston University, Boston, MA 02215
School of Public Communication
Donis Dondis, dean

Bowling Green State University, Bowling Green, OH 43403
School of Journalism
Hal Fisher, director

California State University, Fresno, CA 93740
Department of Journalism
James B. Tucker, chairman

California State University, Fullerton, CA 92634
Department of Communications
Kenward L. Atkin, chairman

California State University, Long Beach, CA 90840
Department of Journalism
Ben Cunningham, chairman

California State University, Northridge, CA 91330
Department of Journalism
Tom Reilly, acting chairman

University of California, Berkeley, CA 94720
Graduate School of Journalism
Edwin R. Bayley, dean

University of Southern California, University Park, Los Angeles, CA 90007
School of Journalism
Nancy E. Wood, director

Colorado State University, Fort Collins, CO 80523
Department of Technical Journalism
David G. Clark, chairman

University of Colorado, Boulder, CO 80309
School of Journalism, Campus Box 287
Russell E. Shain, dean

Columbia University, New York, NY 10027
Graduate School of Journalism
Osborn Elliott, dean

Drake University, Des Moines, IA 50311
School of Journalism and Mass Communication
Herbert Strentz, dean

University of Florida, Gainesville, FL 32611
College of Journalism and Communications
Ralph Lowenstein, dean

University of South Florida, Tampa, FL 33620
Department of Mass Communications
Emery L. Sasser, chairman

University of Georgia, Athens, GA 30602
Henry W. Grady School of Journalism and Mass Communication
Scott M. Cutlip, dean

University of Hawaii at Manoa, Honolulu, HI 96822
Journalism Program
John Luter, director

University of Illinois at Urbana-Champaign, Urbana, IL 61801
College of Communications
James W. Carey, dean

Northern Illinois University, DeKalb, IL 60115
Department of Journalism
Irvan Kummerfeldt, chairman

Southern Illinois University, Carbondale, IL 62901
School of Journalism
Vernon Stone, director

Southern Illinois University, Edwardsville, IL 62026
Department of Mass Communications
John A. Regnell, chairman

Indiana University, Bloomington, IN 47405
School of Journalism
Richard G. Gray, dean

Iowa State University, Ames, IA 50011
Department of Journalism and Mass Communication
J. K. Hvistendahl, chairman

University of Iowa, Iowa City, IA 52242
School of Journalism and Mass Communication
Kenneth Starck, director

Kansas State University, Manhattan, KS 66506
Department of Journalism and Mass Communication
Harry Marsh, head

University of Kansas, Lawrence, KS 66045
William Allen White School of Journalism
Del Brinkman, dean

Kent State University, Kent, OH 44242
School of Journalism
Ralph C. Darrow, acting director

University of Kentucky, Lexington, KY 40506
School of Journalism
Robert D. Murphy, director

Western Kentucky University, Bowling Green, KY 42101
Department of Journalism
David B. Whitaker, head

Louisiana State University, University Station, Baton Rouge, LA 70803
School of Journalism
John C. Merrill, director

Marshall University, Huntington, WV 25701
W. Page Pitt School of Journalism
Deryl R. Leaming, director

University of Maryland, College Park, MD 20742
College of Journalism
Reese Cleghorn, dean

Memphis State University, Memphis, TN 38152
Department of Journalism
Gerald C. Stone, chairman

Michigan State University, East Lansing, MI 48824
College of Communication Arts and Sciences
Erwin P. Bettinghaus, dean

University of Michigan, Ann Arbor, MI 48109
Department of Communication
Peter Clarke, chairman

University of Minnesota, Minneapolis, MN 55455
School of Journalism and Mass Communication
F. Gerald Kline, director

University of Mississippi, University, MS 38677
Department of Journalism
H. W. Norton, Jr., chairman

University of Missouri, Columbia, MO 65205
School of Journalism, Box 838
Roy M. Fisher, dean

University of Montana, Missoula, MT 59812
School of Journalism
Warren J. Brier, dean

University of Nebraska–Lincoln, NE 68588
School of Journalism
Neale Copple, dean

University of Nevada, Reno, NV 89557
Department of Journalism
Robert Blair Kaiser, chairman

University of New Mexico, Albuquerque, NM 87131
Department of Journalism
Robert H. Lawrence, chairman

New York University, New York, NY 10003
Department of Journalism and Mass Communication
David M. Rubin, chairman

University of North Carolina, Chapel Hill, NC 27514
School of Journalism
Richard R. Cole, dean

University of North Dakota, Grand Forks, ND 58202
Department of Journalism
Vern Keel, chairman

Northwestern University, Evanston, IL 60201
Medille School of Journalism
I. W. Cole, dean

Ohio State University, Columbus, OH 43210
School of Journalism
Walter Bunge, director

Ohio University, Athens, OH 45701
Cortland Anderson, director

Oklahoma State University, Stillwater, OK 74078
School of Journalism and Broadcasting
Harry E. Heath, director

University of Oklahoma, Norman, OK 73019
H. H. Herbert School of Journalism and Mass Communication
Elden E. Rawlings, director

Oregon State University, Corvallis, OR 97331
Department of Journalism
Fred C. Zwahlen, chairman

University of Oregon, Eugene, OR 97401
School of Journalism
Everette E. Dennis, dean

Pennsylvania State University, University Park, PA 16802
School of Journalism
Robert O. Blanchard, director

San Jose State University, San Jose, CA 95192
Department of Journalism and Mass Communications
Dennis E. Brown, chairman

University of South Carolina, Columbia, SC 29208
College of Journalism
Albert T. Scroggins, dean

South Dakota State University, Brookings, SD 57006
Department of Journalism and Mass Communication
Richard W. Lee, head

Syracuse University, Syracuse, NY 13210
S. I. Newhouse School of Public Communications
Edward C. Stephens, dean

Temple University, Philadelphia, PA 19122
Department of Journalism
Paul W. Sullivan, chairman

University of Tennessee, Knoxville, TN 37916
College of Communications
Donald G. Hileman, dean

Texas A & M University, College Station, TX 77843
Department of Communications
Bob G. Rogers, head

Texas Christian University, Ft. Worth, TX 76129
Department of Journalism
Doug Newsom, chairman

North Texas State University, Denton, TX 76203
Department of Journalism
Reg Westmoreland, chairman

Texas Tech University, Lubbock, TX 79409
Department of Mass Communications
Billy I. Ross, chairman

University of Texas, Austin, TX 78712
College of Communication
Robert C. Jeffrey, dean

University of Utah, Salt Lake City, UT 84112
Department of Communication
Robert K. Tiemens, chairman

Virginia Commonwealth University, Richmond, VA 23284
Department of Mass Communications
George T. Crutchfield, chairman

University of Washington, Seattle, WA 98195
School of Communications
Don R. Pember, director

Washington and Lee University, Lexington, VA 24450
Lee Memorial Journalism Foundation
R. H. MacDonald, director

West Virginia University, Morgantown, WV 26506
Perley Isaac Reed School of Journalism
Guy H. Stewart, dean

University of Wisconsin, Eau Claire, WI 54701
Department of Journalism
Chairman

University of Wisconsin, Madison, WI 53706
School of Journalism and Mass Communication
James L. Hoyt, acting director

University of Wisconsin, Milwaukee, WI 53201
Department of Mass Communication
Earl Grow, chairman

University of Wisconsin, Oshkosh, WI 54901
Department of Journalism
David J. Lippert, chairman

APPENDIX B

The following is an outline of the undergraduate requirements for a degree in journalism at the University of Iowa, Iowa City, Iowa. A student may seek a journalism degree as a bachelor of arts or as a bachelor of sciences, depending upon his interests. The University of Iowa program is typical of those that combine a double major program. In the Iowa program, a student seeking a broadcast journalism career would probably select the *Mass Communication Laboratory Sequence.* The University's other two sequences are provided to demonstrate the varied courses of study from which a student may select in the sequences.

JOURNALISM AT IOWA
The University of Iowa
School of Journalism and Mass Communication
UNDERGRADUATE PROGRAM*

An Introduction to Iowa Journalism and Mass Communication

Our main objective in the undergraduate program is to prepare students for professional positions in journalism and for careers in the broad field of mass communication. Such positions vary widely but include newspaper reporting and editing, magazine writing and editing, broadcast journalism, public relations, organizational communication, book publication, media graphics and design and photography. The Iowa program emphasizes the basics of reporting, writing and editing. But professional preparation also requires an introduction to and an understanding of theoretical concepts. In all courses, we strive to achieve an integration of practice and theory. Our program offers a wide variety of courses.

*This information sheet is primarily for undergraduate journalism students at Iowa. For additional information, write or call the School of Journalism and Mass Communication, the University of Iowa, Iowa City, Iowa 52242 (Telephone: 319-353-5414). For information about the School's master's or doctoral programs, please request a copy of the "Graduate Studies Handbook."

3/82

We strongly believe students should have a strong liberal arts background to go with their professional preparation. This is preparation for life-long learning. Thus, we limit students to 36 semester hours in the School of Journalism and Mass Communication. From course work taken outside the journalism program, students must develop an area of concentration, a second major, or equivalent, in consultation with their advisers. To complete this requirement, students may either (1) complete the major requirements of another department, or (2) create their own area of concentration by selecting related courses in several departments for a total of 24–30 semester hours of credit beyond the core level.

The Iowa program offers undergraduates a choice of three sequences of study: News-Editorial, Mass Communication Laboratory and Mass Communication Inquiry. Students in all sequences must fulfill the following foundation requirements:

Required Courses for All Journalism Majors—Foundation Courses

—19:101	Cultural and Historical Foundations of Communications	3 s.h.
—19:103	Social Scientific Foundations of Communication	3 s.h.
—19:110	Introduction to Journalistic Writing	3 s.h.
—19:130	Legal and Ethical Issues in Communication	3 s.h.
		12 s.h.

After completing the 12 semester hours of foundation courses, students select one of three sequences outlined to fulfill the 30-semester hour degree requirement. During the final semester, all graduating seniors are required to take 19:185, Contemporary Issues and Problems in Mass Communication, for one semester hour of credit.

News-Editorial Sequence

This sequence is concerned with the gathering, organizing and effective writing of news and other information from printed, human and environmental sources. It also involves the processing, packaging and display of news stories, articles and illustrations for printed and broadcast media. Courses provide opportunities for the development of the various technical skills required for work in the student's choice of media. Career possibilities for students in this sequence include daily or community newspapers, magazines, broadcast journalism, public relations and other professional positions in the news media. The sequence is accredited by the Accrediting Council on Education in Journalism and Mass Communication (ACEJMC). These are the required journalism courses:

—Foundation courses	12 s.h.
—19:112 News Reporting and Writing	3 s.h.
—19:114 News Processing	3 s.h.
—19:116 Advanced Reporting	3 s.h.
—Journalism electives	8 s.h.
—19:185 Contemporary Issues and Problems in Mass Communication	1 s.h.
TOTAL REQUIRED	30 s.h.

Maximum journalism credits allowed toward graduation: 36 s.h.

Mass Communication Laboratory Sequence

This sequence offers students an opportunity to develop proficiency as professional communicators who can identify and analyze problems that need communication strategies and media products for solutions. Students in this sequence combine reporting, production and conceptual courses within the context of their intellectual and media interests. Seniors in the Mass Communication Lab are formed into enterprise teams that develop independent productions or projects for clients in need of professional communication services. These projects may include the development of slide/tape presentations, videotape productions, brochures and other publications. Career possibilities for students completing the sequence include working for public relations departments, advertising agencies, public information offices, independent production companies, as well as either print or broadcast journalism. These are the required journalism courses:

—Foundation courses	12 s.h.
—One reporting course, selected from:	3 s.h.
19:112 News Reporting and Writing	
19:135 Broadcast Journalism	
—One production course, selected from:	3 s.h.
19:136 Broadcast Journalism Workshop	
19:150 Photocommunication I	
19:166 Graphic Design and Production	
—19:181 Mass Communication Lab	3 s.h.
—Journalism electives	8 s.h.
—19:185 Contemporary Issues and Problems in Mass Communication	1 s.h.
TOTAL REQUIRED	30 s.h.

Maximum journalism credits allowed toward graduation: 36 s.h.

Mass Communication Inquiry Sequence

This sequence emphasizes the acquisition of knowledge about communication and concentrates on the study of communication as a way of comprehending society and human interaction. Students take courses which focus on historical, philosophical and social scientific modes of understanding. Career possibilities for students in this sequence include public relations, media research and public opinion polling or other related careers. Many students will continue on to graduate studies in journalism or mass communication or in other disciplines. These are the required journalism courses:

—Foundation courses	12 s.h.
—19:174 Communication Research Methods	3 s.h.
—One course, selected from:	3 s.h.
19:155 Communication and Public Relations	
19:178 Mass Media and Society	
—19:182 Special Topics in Communication	3 s.h.
—Journalism electives	8 s.h.
—19:185 Contemporary Issues and Problems in Mass Communication	1 s.h.
TOTAL REQUIRED	30 s.h.

Maximum journalism credits allowed toward graduation: 36 s.h.

TWO DEGREE PROGRAMS—B.A. AND B.S. DEGREES

The School offers the B.A. and B.S. degrees. These are the requirements:

B.A. REQUIREMENTS:

—Four semesters of a foreign language (or equivalent).
—Foundation Courses.
—Sequence Courses.
—19:185, Contemporary Issues and Problems in Mass Communication.
—Fulfillment of the J-School's second area of concentration requirement in one of two ways:

1. A full B.A. major in another department.
2. A 24–30 semester hour concentration beyond the core level. This concentration should be designed by the student and approved in advance by the student's adviser.

B.S. REQUIREMENTS:

—Two semesters of a foreign language (or equivalent).
—Foundation Courses.
—Sequence Courses.
—19:185, Contemporary Issues and Problems in Mass Communication.
—Six semester hours of natural or social science methods courses.
—Fulfillment of the J-School's second area of concentration requirement in one of two ways:

1. A full B.S. major in a natural or social science.
2. A 24–30 semester hour concentration in the natural or social sciences beyond the core level. This concentration should be designed by the student and approved in advance by the student's adviser.

Honors Program

Freshmen and upperclassmen with outstanding academic records may participate in the Honors Program. They are urged to see the J-School's Honors Program adviser as soon as possible. After admission to the Honors Program, a student must fulfill these requirements:

—Carry out additional work under the guidance of an instructor in the context of any one of the advanced conceptual courses in journalism or mass communication.
—Enroll in 19:196, Honors Colloquium, 3 s.h.
—Write an honors thesis under the supervision of a faculty member.
—Make a formal presentation of honors work to a committee consisting of a faculty adviser, the coordinator of the journalism honors program and a third faculty member of the student's choice.

Minor in Journalism

To meet the requirements for a minor in journalism and mass communication, a student must complete at least 16 semester hours in journalism and mass communication, 12 of which must be in the following courses:

19:101	Cultural and Historical Foundations of Communication	3 s.h.
19:103	Social Scientific Foundations of Communication	3 s.h.
19:110	Introduction to Journalistic Writing	3 s.h.
19:130	Legal and Ethical Issues in Communication	3 s.h.

Transfer work in introductory courses will be considered toward the minor but must be approved by a J-School adviser. No courses for the minor requirement may be taken pass-nonpass. A student must have at least a 2.0 grade-point average in the minor courses. At the time they apply for a degree, students must inform the Office of the Registrar of their desire to have a minor listed on their transcript.

Transfer Students

The J-School's policy is to accept journalism transfer credits from another institution of up to but not more than 20 percent of the student's total number of credits toward a major in journalism at Iowa. Other course work taken elsewhere might be applicable toward fulfilling elective and/or second area of concentration requirements. Any transfer credit intended to meet School of Journalism and Mass Communication requirements must, of course, be approved by the student's journalism adviser at Iowa.

APPENDIX C

The following are free booklets and pamphlets on broadcast journalism or other journalism careers that will be useful for a student who is exploring or considering a journalism career.

Careers in Radio
Publications Department
National Association of Broadcasters
1771 N Street, NW
Washington, DC 20036

Careers in Broadcast News
Radio-Television News Directors Association
1735 DeSales Street, NW
Washington, DC 20036

Careers in Television
Publications Department
National Association of Broadcasters
1771 N Street, NW
Washington, DC 20036

Careers in Communications
Women in Communication
P.O. Box 9561
Austin, TX 78766

Business Communication as a Career
International Association of Business Communicators
Suite 940, 870 Market Street
San Francisco, CA 94102

Getting Started in Writing
Writer's Digest
9933 Alliance Road
Cincinnati, OH 45252

C'Mon Up to a Rewarding Career in Journalism
National Newspaper Association
1627 K Street, NW, Suite 400
Washington, DC 20006

Today's Journalism Students: Who They Are and What They Want to Do
The Gannett Foundation
Lincoln Tower
Rochester, NY 14604

Newspaper Jobs You Never Thought Of . . . Or Did You?
American Newspaper Publishers Association Foundation
P.O. Box 17407
Dulles International Airport
Washington, DC 20041

Newspaper Jobs for Journalism Grads
American Newspaper Publishers Association Foundation
P.O. Box 17407
Dulles International Airport
Washington, DC 20041

Your Future in Daily Newspapers
American Newspaper Publishers Association Foundation
P.O. Box 17407
Dulles International Airport
Washington, DC 20041

Scholarship Guide Book

1982 Journalism Scholarship Guide and Directory of College Journalism Programs. The Newspaper Fund, P.O. Box 300, Princeton, NJ 08540.

APPENDIX D

Sources of Information on Broadcasting and Related Areas

Action for Children's Television
46 Austin Street
Newtonville, MA 02160

Advertising Council
825 Third Avenue
New York, NY 10022

Advertising Research Foundation, Inc.
3 East 54th Street
New York, NY 10022

Affiliated Advertising Agencies International
516 Fifth Avenue
New York, NY 10036

Alpha Epsilon Rho
(National Honorary Broadcasting Society)
College of Journalism
University of South Carolina
Columbia, SC 29208

American Advertising Federation, Inc.
1225 Connecticut Avenue NW
Washington, DC 20036

American Association of Advertising Agencies
200 Park Avenue
New York, NY 10017

American Council for Better Broadcasts
120 East Wilson Street
Milwaukee, WI 53702

American Women in Radio and Television
1321 Connecticut Avenue NW
Washington, DC 20036

Associated Press Broadcasters
50 Rockefeller Center
New York, NY 10020

Association of Independent Television Stations, Inc.
19 West 44th Street
New York, NY 10036

Association of Public Radio Stations
1730 Pennsylvania Avenue
Washington, DC 20006

Corporation for Public Broadcasting
1111 16th Street NW
Washington, DC 20036

Division of Radio-Television of the Association for Education in Journalism
206 Avery Hall
University of Nebraska
Lincoln, NE 68508

Independent Television News Association
225 East 42nd Street
New York, NY 10017

National Association of Broadcasters
1771 N Street NW
Washington, DC 20036

National Association of Educational Broadcasters
1346 Connecticut Avenue NW
Washington, DC 20036

National Association of Farm Directors
KGNC
Amarillo, Texas

National Association of Television and Radio Announcers
1408 South Michigan Avenue
Chicago, IL 60605

National Broadcast Association for Community Affairs
4400 Jenifer Street NW
Washington, DC 20015

National Broadcast Editorial Association
514 West 57th Street
New York, NY 10019

National News Council
1 Lincoln Plaza
New York, NY 10023

National Radio Broadcasters Association
500 Fifth Avenue, Suite 1450
New York, NY 10036

Radio-Television News Directors Association
1735 DeSales Street NW
Washington, DC 20510

Radio-TV Correspondents' Association
The Capitol, S 312
Washington, DC 20510

Radio-Television News Directors of Canada
Box 370
Sault Ste. Marie
Ontario, Canada

Sigma Delta Chi
Professional Journalistic Society
35 East Wacker Drive
Chicago, IL 60601

Television Information Office
745 Fifth Avenue
New York, NY 10022

Women in Communications, Inc.
8305-A Shoal Creek Boulevard
Austin, TX 78758

RECOMMENDED READING

Air Time: The Inside Story of CBS News, Gary Paul Gates. New York: Harper & Row.

Before the Color Fades, Harry Reasoner. New York: Random House.

Careers in Journalism for the New Woman, Megan Rosenfeld. New York: Watts.

Clearing the Air, Daniel Schorr. Boston: Houghton Mifflin.

Community Journalism, Kenneth R. Byerly. New York: Chilton.

Deciding What's News: A Study of CBS News, Herbert J. Gans. New York: Pantheon Books.

Do You Belong in Journalism?, Henry Gemmill and Bernard Kilgore (eds.). New York: Appleton-Century-Crofts.

Exploring Careers in Journalism, Thomas Pawlick. New York: Richards Rosen Press.

Forgive Us Our Press Passes, Elaine Shepard. Englewood, New Jersey: Prentice-Hall.

How to Talk with Practically Anybody About Practically Anything, Barbara Walters. New York: Dell.

Inside Story, Brit Hume. Garden City, New York: Doubleday.

Journalists in Action, Edward W. Barrett (ed.). New York: Channel Press.

Journalist—Eyewitness to History, Herbert Brucker. New York: Macmillan.

Let's Be Frank About It, Frank Blair. Garden City, New York: Doubleday.

Like It Is, Howard Cosell. Chicago: Playboy Press.

News from Nowhere, Edward Jay Epstein. New York: Random House.

Opportunities in Broadcasting, Elmo Ellis. Skokie, Illinois: VGM Career Horizons.

Prime Time: The Life of Edward R. Murrow, Alexander Kendrick. Boston: Little, Brown & Company.

Radio News Writing, William Fern Brooks. New York: McGraw-Hill.

So Long Until Tomorrow, Lowell Thomas. New York: William Morrow.

Star Reporters and 34 of Their Greatest Stories, Ward Green (ed.). New York: Random House.

Talking Woman, Shana Alexander. New York: Delacorte Press.

The Camera Never Blinks, Dan Rather. New York: William Morrow.
The Good Guys, the Bad Guys and the First Amendment, Fred Friendly. New York: Random House.
The Newscasters, Ron Powers. New York: St. Martin's Press.
The Palace Guard, Dan Rather and Gary Paul Gates. New York: Harper & Row.
The Press in Washington, Ray Hiebert. New York: Dodd, Mead.
Those Radio Commentators, Irving E. Fang. Ames: Iowa State University Press.
Television and American Culture, Carl Lowe. New York: H.W. Wilson Company.
Television News Reporting, CBS News Staff. New York: McGraw-Hill.
Television and Radio, Giraud Chester, Garnet R. Garrison, and Edgar Willis. New York: Appleton-Century-Crofts.
Television and Radio News, Bob Siller. New York: Macmillan.
Television News, Irving E. Fang. New York: Hastings House.
Women in Television News, Judith S. Gelfman. New York: Columbia University Press.
Your Career in Journalism, Meyer L. Stein. New York: Messner.